T0350149

Materials and Devices for Laser Remote Sensing and Optical Communication

MATERIALS RESEARCH SOCIETY
SYMPOSIUM PROCEEDINGS VOLUME 1076

Materials and Devices for Laser Remote Sensing and Optical Communication

Symposium held March 25–27, 2008, San Francisco, California, U.S.A.

EDITORS:

Astrid Aksnes

Norwegian University of Science and Technology
Department of Electronics and Telecommunications
Trondheim, Norway

Farzin Amzajerdian

NASA Langley Research Center
Hampton, Virginia, U.S.A.

Materials Research Society
Warrendale, Pennsylvania

CAMBRIDGE
UNIVERSITY PRESS

University Printing House, Cambridge CB2 8BS, United Kingdom

One Liberty Plaza, 20th Floor, New York, NY 10006, USA

477 Williamstown Road, Port Melbourne, VIC 3207, Australia

314-321, 3rd Floor, Plot 3, Splendor Forum, Jasola District Centre, New Delhi - 110025, India

79 Anson Road, #06-04/06, Singapore 079906

Cambridge University Press is part of the University of Cambridge.

It furthers the University's mission by disseminating knowledge in the pursuit of
education, learning and research at the highest international levels of excellence.

www.cambridge.org
Information on this title: www.cambridge.org/9781605110462

Materials Research Society
506 Keystone Drive, Warrendale, PA 15086
http://www.mrs.org

© Materials Research Society 2008

First published 2008
First paperback edition 2012

Single article reprints from this publication are available through
University Microfilms Inc., 300 North Zeeb Road, Ann Arbor, MI 48106

CODEN: MRSPDH

A catalogue record for this publication is available from the British Library

ISBN 978-1-605-11046-2 Hardback
ISBN 978-1-107-40852-4 Paperback

Cambridge University Press has no responsibility for the persistence or
accuracy of URLs for external or third-party internet websites referred to in
this publication, and does not guarantee that any content on such websites is,
or will remain, accurate or appropriate.

These proceedings were sponsored in part by the National Aeronautics and Space Administration
under Cooperative Agreement Number NNX07AV61 A.

This material is based upon work supported in part by the National Science Foundation under
Grant Number ECCS-0802422.

Any opinions, findings, and conclusions or recommendations expressed in this material are
those of the author(s) and do not necessarily reflect the views of the National Aeronautics and
Space Administration and the National Science Foundation.

CONTENTS

Preface ix

Materials Research Society Symposium Proceedings xi

LASER REMOTE SENSING INSTRUMENTS

* Role of Lidar Technology in Future NASA Space Missions 3
   Farzin Amzajerdian

* ALADIN Doppler Wind Lidar and Related Programs at
EADS Astrium 9
   Didier Morancais, Frédéric Fabre, and Yves Toulemont

Geospatial Remote Sensing Using Advanced Sensor Systems 21
   A. Vaseashta

Development Effort of the Lunar Orbiter Laser Altimeter
Laser Transmitter 27
   George B. Shaw, Anthony W. Yu, and
   Anne-Marie D. Novo-Gradac

* All Fiber Coherent Doppler LIDAR for Wind Sensing 35
   Toshiyuki Ando, Shumpei Kameyama, Kimio Asaka,
   Yoshihito Hirano, Hisamichi Tanaka, and
   Hamaki Inokuchi

* Linear FMCW Laser Radar for Precision Range and
Vector Velocity Measurements 47
   Diego F. Pierrottet, Farzin Amzajerdian, Larry Petway,
   Bruce Barnes, George Lockard, and Manuel Rubio

Simplified Homodyne Detection for Linear FM Lidar 57
   Peter Adany, Chris Allen, and Rongqing Hui

FIBER OPTIC AND SEMICONDUCTOR LASERS

* Microstructured Soft Glass Fibers for Advanced Fiber Lasers 65
   Axel Schulzgen, Li Li, Xiushan Zhu, Shigeru Suzuki,
   Valery L. Temyanko, Jacques Albert, and
   Nasser Peyghambarian

*Invited Paper

Efficient High Power ns Pulsed Fiber Laser for Lidar and Laser Communications 75
Jian Liu

* **Electrically Pumped Photonic Crystal Distributed Feedback Quantum Cascade Lasers** 81
Manijeh Razeghi, Yanbo Bai, Steven Slivken, and Wei Zhang

* **Development of Low-Cost Multi-Watt Yellow Lasers Using InGaAs/GaAs Vertical External-Cavity Surface-Emitting Lasers** 91
Mahmoud Fallahi, Li Fan, Chris Hessenius, Jorg Hader, Hongbo Li, Jerome Moloney, Wolfgang Stolz, Stephan Koch, and James Murray

* **Quantum Design of Active Semiconductor Materials for Targeted Wavelengths** 97
Jerome Moloney, Joerg Hader, and Stephan W. Koch

Advanced Laser Diode Cooling Concepts 119
Ryan Feeler, Jeremy Junghans, Edward Stephens, Greg Kemner, Fred Barlow, Jared Wood, and Aicha Elshabini

PHOTODETECTION DEVICES

Single Carrier Initiated Low Excess Noise Mid-Wavelength Infrared Avalanche Photodiode Using InAs-GaSb Strained Layer Superlattice 127
Koushik Banerjee, Shubhrangshu Mallick, Siddhartha Ghosh, Elena Plis, Jean Baptiste Rodriguez, Sanjay Krishna, and Christoph Grein

* **Quantum Well and Quantum Dot Based Detector Arrays for Infrared Imaging** 133
Sarath Gunapala, Sumith Bandara, Cory Hill, David Ting, John Liu, Jason Mumolo, Sam Keo, and Edward Blazejewski

*Invited Paper

* **Infrared Detctor Activities at NASA Langley Research Center** 147
    M. Nurul Abedin, Tamer F. Refaat, Oleg V. Sulima,
    and Farzin Amzajerdian

Structural and Photoconducting Properties of MBE and MOCVD Grown III-Nitride Double-Heterostructures 155
    Sindy Hauguth-Frank, Vadim Lebedev, Katja Tonisch,
    Henry Romanus, Thomas Kups, Hans-Joachim Büchner,
    Gerd Jäger, Oliver Ambacher, and Andreas Schober

NANOCRYSTAL AND PHOTONIC STRUCTURE AND DEVICES

A Buried Silicon Nanocrystals Based High Gain Coefficient $SiO_2/SiO_x/SiO_2$ Strip-Loaded Waveguide Amplifier on Quartz Substrate 165
    Cheng-Wei Lian and Gong-Ru Lin

Surface Modification of ZrO_2 Nanoparticles as Functional Component in Optical Nanocomposite Devices 175
    Ninjbadgar Tsedev and Georg Garnweitner

Temperature Dependent Fluorescence of Nanocrystalline Ce-Doped Garnets for Use as Thermographic Phosphors 181
    Rachael Hansel, Steve Allison, and Greg Walker

Non-Selective Optical Wavelength-Division Multiplexing Devices Based on a-SiC:H Multilayer Heterostuctures 187
    Manuela Vieira, Miguel Fernandes, Paula Louro,
    Manuel Augusto Vieira, Manuel Barata, and
    Alessandro Fantoni

Tailoring Quantum Dot Saturable Absorber Mirrors for Ultra-Short Pulse Generation 193
    Matthew Lumb, Edmund Clarke, Dominic Farrell,
    Michael Damzen, and Ray Murray

Author Index 199

Subject Index 201

*Invited Paper

PREFACE

This volume includes a selection of the papers presented at Symposium K, "Materials and Devices for Laser Remote Sensing and Optical Communication," held March 25–27 at the 2008 MRS Spring Meeting in San Francisco, California. Although laser remote sensing and optical communication are distinct in their application areas and marketplace, they share many common technology elements such as lasers, detectors, modulators, and other photonic and semiconductor devices. This symposium highlighted the advances in these component technology areas and their impact on both laser remote sensing and optical communication applications. In addition, several laser remote sensing concepts benefiting from the advances made by the data and telecommunication industries were presented and discussed. It is hoped that the potential utilization of relevant optical communication technologies for laser remote sensing applications discussed in this symposium will have a meaningful impact on future technology programs.

We wish to thank the symposium co-organizer Nasser Peyghambarian for his important contributions in arranging a series of excellent cross-cutting talks and organizing sessions that gathered technologists from different communities. We also thank all of the contributors and participants who made this symposium successful, especially the invited speakers for their outstanding presentations.

We gratefully acknowledge the financial support provided by:

        IPG Photonics
        PolarOnyx
        Coherent Applications
        Discovery Semiconductors
        National Science Foundation
        National Aeronautic and Space Administration

Astrid Aksnes
Farzin Amzajerdian

July 2008

MATERIALS RESEARCH SOCIETY SYMPOSIUM PROCEEDINGS

Volume 1066 — Amorphous and Polycrystalline Thin-Film Silicon Science and Technology—2008, A. Nathan, J. Yang, S. Miyazaki, H. Hou, A. Flewitt, 2008, ISBN 978-1-60511-036-3

Volume 1067E —Materials and Devices for "Beyond CMOS" Scaling, S. Ramanathan, 2008, ISBN 978-1-60511-037-0

Volume 1068 — Advances in GaN, GaAs, SiC and Related Alloys on Silicon Substrates, T. Li, J. Redwing, M. Mastro, E.L. Piner, A. Dadgar, 2008, ISBN 978-1-60511-038-7

Volume 1069 — Silicon Carbide 2008—Materials, Processing and Devices, A. Powell, M. Dudley, C.M. Johnson, S-H. Ryu, 2008, ISBN 978-1-60511-039-4

Volume 1070 — Doping Engineering for Front-End Processing, B.J. Pawlak, M. Law, K. Suguro, M.L. Pelaz, 2008, ISBN 978-1-60511-040-0

Volume 1071 — Materials Science and Technology for Nonvolatile Memories, O. Auciello, D. Wouters, S. Soss, S. Hong, 2008, ISBN 978-1-60511-041-7

Volume 1072E —Phase-Change Materials for Reconfigurable Electronics and Memory Applications, S. Raoux, A.H. Edwards, M. Wuttig, P.J. Fons, P.C. Taylor, 2008, ISBN 978-1-60511-042-4

Volume 1073E —Materials Science of High-k Dielectric Stacks—From Fundamentals to Technology, L. Pantisano, E. Gusev, M. Green, M. Niwa, 2008, ISBN 978-1-60511-043-1

Volume 1074E —Synthesis and Metrology of Nanoscale Oxides and Thin Films, V. Craciun, D. Kumar, S.J. Pennycook, K.K. Singh, 2008, ISBN 978-1-60511-044-8

Volume 1075E —Passive and Electromechanical Materials and Integration, Y.S. Cho, H.A.C. Tilmans, T. Tsurumi, G.K. Fedder, 2008, ISBN 978-1-60511-045-5

Volume 1076 — Materials and Devices for Laser Remote Sensing and Optical Communication, A. Aksnes, F. Amzajerdian, 2008, ISBN 978-1-60511-046-2

Volume 1077E —Functional Plasmonics and Nanophotonics, S. Maier, 2008, ISBN 978-1-60511-047-9

Volume 1078E —Materials and Technology for Flexible, Conformable and Stretchable Sensors and Transistors, 2008, ISBN 978-1-60511-048-6

Volume 1079E —Materials and Processes for Advanced Interconnects for Microelectronics, J. Gambino, S. Ogawa, C.L. Gan, Z. Tokei, 2008, ISBN 978-1-60511-049-3

Volume 1080E —Semiconductor Nanowires—Growth, Physics, Devices and Applications, H. Riel, T. Kamins, H. Fan, S. Fischer, C. Thelander, 2008, ISBN 978-1-60511-050-9

Volume 1081E —Carbon Nanotubes and Related Low-Dimensional Materials, L-C. Chen, J. Robertson, Z.L. Wang, D.B. Geohegan, 2008, ISBN 978-1-60511-051-6

Volume 1082E —Ionic Liquids in Materials Synthesis and Application, H. Yang, G.A. Baker, J.S Wilkes, 2008, ISBN 978-1-60511-052-3

Volume 1083E —Coupled Mechanical, Electrical and Thermal Behaviors of Nanomaterials, L. Shi, M. Zhou, M-F. Yu, V. Tomar, 2008, ISBN 978-1-60511-053-0

Volume 1084E —Weak Interaction Phenomena—Modeling and Simulation from First Principles, E. Schwegler, 2008, ISBN 978-1-60511-054-7

Volume 1085E —Nanoscale Tribology—Impact for Materials and Devices, Y. Ando, R.W. Carpick, R. Bennewitz, W.G. Sawyer, 2008, ISBN 978-1-60511-055-4

Volume 1086E —Mechanics of Nanoscale Materials, C. Friesen, R.C. Cammarata, A. Hodge, O.L. Warren, 2008, ISBN 978-1-60511-056-1

MATERIALS RESEARCH SOCIETY SYMPOSIUM PROCEEDINGS

Volume 1087E —Crystal-Shape Control and Shape-Dependent Properties—Methods, Mechanism, Theory and Simulation, K-S. Choi, A.S. Barnard, D.J. Srolovitz, H. Xu, 2008, ISBN 978-1-60511-057-8

Volume 1088E —Advances and Applications of Surface Electron Microscopy, D.L. Adler, E. Bauer, G.L. Kellogg, A. Scholl, 2008, ISBN 978-1-60511-058-5

Volume 1089E —Focused Ion Beams for Materials Characterization and Micromachining, L. Holzer, M.D. Uchic, C. Volkert, A. Minor, 2008, ISBN 978-1-60511-059-2

Volume 1090E —Materials Structures—The Nabarro Legacy, P. Müllner, S. Sant, 2008, ISBN 978-1-60511-060-8

Volume 1091E —Conjugated Organic Materials—Synthesis, Structure, Device and Applications, Z. Bao, J. Locklin, W. You, J. Li, 2008, ISBN 978-1-60511-061-5

Volume 1092E —Signal Transduction Across the Biology-Technology Interface, K. Plaxco, T. Tarasow, M. Berggren, A. Dodabalapur, 2008, ISBN 978-1-60511-062-2

Volume 1093E —Designer Biointerfaces, E. Chaikof, A. Chilkoti, J. Elisseeff, J. Lahann, 2008, ISBN 978-1-60511-063-9

Volume 1094E —From Biological Materials to Biomimetic Material Synthesis, N. Kröger, R. Qiu, R. Naik, D. Kaplan, 2008, ISBN 978-1-60511-064-6

Volume 1095E —Responsive Biomaterials for Biomedical Applications, J. Cheng, A. Khademhosseini, H-Q. Mao, M. Stevens, C. Wang, 2008, ISBN 978-1-60511-065-3

Volume 1096E —Molecular Motors, Nanomachines and Active Nanostructures, H. Hess, A. Flood, H. Linke, A.J. Turberfield, 2008, ISBN 978-1-60511-066-0

Volume 1097E —Mechanical Behavior of Biological Materials and Biomaterials, J. Zhou, A.G. Checa, O.O. Popoola, E.D. Rekow, 2008, ISBN 978-1-60511-067-7

Volume 1098E —The Hydrogen Economy, A. Dillon, C. Moen, B. Choudhury, J. Keller, 2008, ISBN 978-1-60511-068-4

Volume 1099E —Heterostructures, Functionalization and Nanoscale Optimization in Superconductivity, T. Aytug, V. Maroni, B. Holzapfel, T. Kiss, X. Li, 2008, ISBN 978-1-60511-069-1

Volume 1100E —Materials Research for Electrical Energy Storage, J.B. Goodenough, H.D. Abruña, M.V. Buchanan, 2008, ISBN 978-1-60511-070-7

Volume 1101E —Light Management in Photovoltaic Devices—Theory and Practice, C. Ballif, R. Ellingson, M. Topic, M. Zeman, 2008, ISBN 978-1-60511-071-4

Volume 1102E —Energy Harvesting—From Fundamentals to Devices, H. Radousky, J. Holbery, B. O'Handley, N. Kioussis, 2008, ISBN 978-1-60511-072-1

Volume 1103E —Health and Environmental Impacts of Nanoscale Materials—Safety by Design, S. Tinkle, 2008, ISBN 978-1-60511-073-8

Volume 1104 — Actinides 2008—Basic Science, Applications and Technology, B. Chung, J. Thompson, D. Shuh, T. Albrecht-Schmitt, T. Gouder, 2008, ISBN 978-1-60511-074-5

Volume 1105E —The Role of Lifelong Education in Nanoscience and Engineering, D. Palma, L. Bell, R. Chang, R. Tomellini, 2008, ISBN 978-1-60511-075-2

Volume 1106E —The Business of Nanotechnology, L. Merhari, A. Gandhi, S. Giordani, L. Tsakalakos, C. Tsamis, 2008, ISBN 978-1-60511-076-9

Volume 1107 — Scientific Basis for Nuclear Waste Management XXXI, W.E. Lee, J.W. Roberts, N.C. Hyatt, R.W. Grimes, 2008, ISBN 978-1-60511-079-0

Prior Materials Research Society Symposium Proceedings available by contacting Materials Research Society

Laser Remote Sensing Instruments

Role of Lidar Technology in Future NASA Space Missions

Farzin Amzajerdian

NASA Langley Research Center, MS 468, Hampton, VA, 23681

ABSTRACT

The past success of lidar instruments in space combined with potentials of laser remote sensing techniques in improving measurements traditionally performed by other instrument technologies and in enabling new measurements have expanded the role of lidar technology in future NASA missions. Compared with passive optical and active radar/microwave instruments, lidar systems produce substantially more accurate and precise data without reliance on natural light sources and with much greater spatial resolution. NASA pursues lidar technology not only as science instruments, providing atmospherics and surface topography data of Earth and other solar system bodies, but also as viable guidance and navigation sensors for space vehicles. This paper summarizes the current NASA lidar missions and describes the lidar systems being considered for deployment in space in the near future.

CURRENT SPACE-BASED LIDARS

Presently, NASA has four major lidar instruments in space with another to be launched later this year. The ICESat (Ice, Cloud and land Elevation Satellite) and CALIPSO (Cloud-Aerosol Lidar and Infrared Pathfinder Satellite Observation) are Earth science missions providing valuable atmospheric data and monitoring global climate changes [1-4]. The other three instruments are part of planetary missions: Mercury Laser Altimeter (MLA) as part of the MESSENGER (MErcury Surface, Space ENvironment, GEochemistry, and Ranging) mission [5], Mars Meteorological Lidar onboard Phoenix Lander and Lunar Orbiter Laser Altimeter (LOLA) onboard the Lunar Reconnaissance Orbiter. Table 1 summarizes the high-level specifications of these instruments and their launch dates. All these instruments utilize diode-pumped Nd:YAG laser as their transmitter source and incorporate some level of redundancy by using backup lasers to ensure long operational lifetime in space.

Table 1. Current NASA Space-based Scientific Lidar Instruments.

Mission	Lidar Instruments	Primary data products	Pulse Energy	Rep rate	No. of Pump Bars	Peak Power/bar	No. of lasers	Telescope Aperture Diameter	Launch date	Required Lifetime In Space
ICESat (Earth Science)	Laser Altimeter w/ atm channel	Ice sheet height, Clouds	110 mJ	40	54	65-85 W	3	100 cm	January 12,2003	3
CALIPSO (Earth Science)	Atmospheric Backscatter	Clouds and aerosol profiles	220 mJ	20	192	50 W	2	100 cm	April 28, 2006	3
MESSENGER	Laser Altimeter	Mercury Surface Topography	20 mJ	8	10	100 W	1	11.5 cm X4	August 3, 2004	7
Phoenix Mars Lander	Atmospheric Backscatter	Mars aerosol, Ice clouds	1 mJ	100			1	10 cm	August 4, 2007	0.2
Lunar Reconnaissance Orbiter	Laser Altimeter	global lunar topographic model	2.7 mJ	28	4	70 W	2	14 cm	Late 2008	2

ICESat, CALIPSO and the Phoenix Meteorological Lidar instruments use external cavity doublers to generate 532-nm radiation along with the fundamental 1064-nm beam. The 532-nm beam profiles the atmospheric aerosols, while the 1064-nm beam is used for denser aerosol plumes, clouds or ground targets. ICESat lidar uses its 1064-nm beam to measure the height of the polar ice sheets and monitor its changes over time. Changes in ice sheet elevations are an important indicator of the global climate change and influence the global sea level, which can have profound impact on life on Earth. CALIPSO uses both laser wavelengths for collecting spatially resolved clouds and aerosols data. The combination of the signal returns at 1064-nm and two orthogonal polarizations of 532-nm radiation allows the scientists to extract information about cloud water and ice content, and aerosol concentration and its sources. The improved information on global coverage of clouds, their altitudes, and their water and ice contents is critical to better weather forecasting and more accurate climate models. The atmospheric aerosols also affect weather and global climate changes. Aerosols can both reflect the sunlight away from Earth, causing the atmosphere to cool and absorb the sunlight, warming the atmosphere depending on aerosol concentration and composition. Figure 1 illustrates the CALIPSO instrument and its measurements of atmospheric aerosols and clouds. CALIPSO payload also includes an Imaging Infrared Radiometer (IIR) and a Wide Field Camera (WFC) to provide additional data for determining the cloud emissivity and ice particle size distribution.

Unlike ICESat and CALIPSO, the Mars Meteorological Lidar onboard Phoenix operates from a stationary platform on the ground looking upward. The Mars Meteorological Lidar measures the atmospheric aerosol concentration and CO_2 ice clouds that can lead to a better understanding of Mars climate and atmospheric processes. The laser altimeter instruments (MLA and LOLA) orbiting Mercury and the moon simply measure the distance to the surface as they orbit. The data collected after a large number of orbits enable the development of global three-dimensional topographical maps.

ICESat and CALIPSO have been in Earth orbit since 2003 and 2006, respectively, and are continuing to operate and transmit data. MESSENGER (MErcury Surface, Space ENvironment, GEochemistry, and Ranging) made its first flyby of Mercury in January 2008, making lidar measurements of the planet's surface. The lidar onboard MESSENGER is expected to operate well beyond 2011 when the spacecraft settles in Mercury's orbit after flybys in October 2008 and September 2009. The Mars Meteorological Lidar is expected to provide clouds, fogs, and dust plumes data leading to be better understanding of the Mars climate after the successful recent landing of Phoenix spacecraft in May 2008.

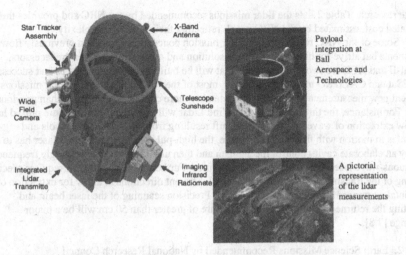

Figure 1. CLAIPSO instrument and a pictorial representation of its measurements.

CANDIDATE LIDAR INSTRUMENTS FOR FUTURE NASA MISSIONS

The lidar technology is now planned for a wide range of applications that can enable NASA's achievement of its scientific and space exploration goals. These applications fall into four general categories:

a) Earth Science: long-duration orbiting instruments providing global monitoring of the atmosphere and land

b) Planetary Science: orbiting or land-based scientific instruments providing geological and atmospheric data of solar system bodies

c) Landing Aid: sensors providing hazard avoidance, guidance and navigation data

d) Rendezvous and Docking Aid: sensors providing spacecraft bearing, distance, and approach velocity

Earth Science Applications

Earth science reliance on lidar technology is clearly revealed in a National Research Council (NRC) report published in 2007, entitled "Earth Science and Applications from Space: National Imperatives for the Next Decade and Beyond" [6]. This report reflects the scientific community's consensus and is regarded very seriously by NASA planners. It provides a list of 15 recommended missions for NASA to implement over the next decade. Seven of the 15 are based on lidar instruments, with the remaining missions divided between passive optical, radar and microwave instrument technologies. This fact reflects the scientific community's belief that the ability of lidar to provide highly accurate atmospheric data on a global scale can profoundly help climate research. Table 2 lists the lidar missions recommended by the NRC and provides their associated cost, estimated by the NRC panel, reflecting their relative complexity.

Some of the NRC recommended lidar mission concepts are based on previously flown instruments but strive for higher accuracy, resolution and coverage than their predecessors. ICESat-II and ACE are two such missions that will be built upon the experiences and successes of ICESat and CALIPSO missions. However, most of the NRC recommended lidar missions represent new measurements requiring considerably more complex instruments than previously flown. For instance, the three-dimensional Wind Lidar will use highly spectral pure pulsed lasers to allow extraction of wavelength Doppler shift resulting from atmospheric aerosols and molecules in motion with the wind. Therefore, the high-pulse energy transmitter laser has to employ an elaborate cavity control mechanism and then use a separate stable, single frequency, continuous-wave laser as a injection-seeding source. Furthermore, a wind lidar requires precision pointing of the transmitted laser beam in several different directions to allow for extraction of the horizontal components of vector wind velocity. Precision scanning of the laser beam and collecting the returned radiation using an aperture of greater than 50 cm will be a major challenge [7,8].

Table 2. Earth Science Missions Recommended by National Research Council

Mission	Mission Description	Timeframe	Instrument
ICESat-II	Ice sheet height changes for climate change diagnosis	2010 to 2013	Laser altimeter
DESDynI	Surface and ice sheet deformation for understanding natural hazards and climate; vegetation structure for ecosystem health		Laser altimeter
ASCENDS	Day/night, all-latitude, all-season CO_2 column integrals for climate emissions	2013 to 2016	Multifrequency laser
ACE	Aerosol and cloud profiles for climate and water cycle; ocean color for open ocean biogeochemistry		Backscatter lidar
LIST	Land surface topography for landslide hazards and water runoff	2016 to 2020	Laser altimeter
GRACE-II	High temporal resolution gravity fields for tracking large-scale water movement		Microwave or laser ranging system
3D-Winds (demo)	Tropospheric winds for weather forecasting and pollution transport		Doppler lidar

Planetary Science Applications

Planetary science applications are mostly focused on the geology and surface topography of the moon, Mars, and other solar system bodies, as well as characterization of the Mars atmosphere. In addition to the missions discussed earlier, NASA will continue deployment of laser altimeters orbiting various solar bodies. Specifically, we are planning to launch the Lunar Orbiter Laser Altimeter (LOLA) later this year. LOLA is expected to orbit the moon for at least two years and provide a three-dimensional surface map of the entire lunar surface. This information will be critical in landing site selection and designing the future robotic and manned landing missions to moon.

The application of lidar technology for understanding the Mars atmosphere is starting to attract the attention of NASA scientists. The NASA report, "Mars Scientific Goals, Objectives, Investigations, and Priorities," prepared in 2006, outlines a series of measurements critical to understanding the Mars atmosphere and search for evidence of life. Many of these measurements – including atmospheric density variations, seasonal and diurnal cycles, aerosol concentration profiles, and detection of water vapor – are best achieved by lidar instruments.

Planetary Landing Applications

Landing aid is another important application of lidar technology in space. Future planetary exploration missions will require safe, precision soft-landing to target scientifically interesting sites near hazardous terrain features, such as escarpments, craters, slopes and rocks. Although the landing accuracy has steadily improved over time to approximately 10 km for Mars landing and 1 km for the moon, a drastically different guidance, navigation and control concept is required to meet future mission requirements. For example, future rovers will require better than 1 km landing accuracy for Mars and better than 30 m for the Moon. Laser radar or lidar technology can be key to meeting these objectives because it can provide high-resolution three-dimensional maps of the terrain, accurate ground proximity and velocity measurements, and it can determine atmospheric pressure and wind velocity in the case of Mars landing [9-11]. These lidar capabilities can enable the landers of the future to identify the preselected landing zone and hazardous terrain features within it, determine the optimum flight path, and accurately navigate using precision ground proximity and velocity data.

Currently, NASA is actively advancing the lidar technology for future lunar landing missions through its Autonomous Landing and Hazard Avoidance Technology (ALHAT) project. This program is developing three-dimensional imaging and Doppler velocity lidar technologies as part of the landing guidance, navigation and control system. The lidar sensors being developed under ALHAT will enable safe soft-landing of large robotic, cargo and crewed vehicles with a high degree of precision at the designated landing site under any lighting conditions.

Rendezvous and Docking Application

The future crew exploration vehicle, which is to replace the space shuttle and be used for a crewed mission to the moon, may rely on a lidar sensor for its rendezvous and docking maneuvers. The lidar technique is being considered for providing critical distance, approach velocity, and relative orientation of the docking port during the rendezvous and docking maneuver. The precision and frequent update rate offered by the lidar could be key for mating the vehicle with the International Space Station and, in the case of the human mission to the moon, for mating the lunar crew module with the Earth re-entry vehicle that will be awaiting it in the moon orbit.

SUMMARY

The lidar technology will play an increasingly important role in NASA's plan. Lidar will be used for a wide range of applications both as a scientific instrument and as a GN&C sensor in many future NASA space missions. Despite past successful lidar missions, deployment of laser systems in space remains a very challenging task. This challenge is further evidenced by the demand for increasing accuracy with higher coverage measurements and the requirement of a number of new measurements leading to considerably complex lidar instruments.

REFERENCES

1. Schutz, B. E., H. J. Zwally, C. A. Shuman, D. Hancock, and J. P. DiMarzio, "Overview of the ICESat Mission," Geophys. Res. Lett., 32, November 2005.
2. Abshire, J. B., X. Sun, H. Riris, J. M. Sirota, J. F. McGarry, S. Palm, D. Yi, and P. Liiva, "Geoscience Laser Altimeter System (GLAS) on the ICESat Mission: On-orbit measurement performance," Geophys. Res. Lett., 32, November 2005.
3. Winker, D. M., J. Pelon, and M. P. McCormick, "The CALIPSO mission: Spaceborne lidar for observation of aerosols and clouds," Proc. SPIE 4893, 1-11, 2003
4. Winker, D. M., W. H. Hunt, and C. A. Hostetler, "Status and Performance of the CALIPSO lidar," Proc. SPIE, 5575, 8-15, 2004.
5. Luis Ramos-Izquierdo, et al., "Optical system design and integration of the Mercury Laser Altimeter," Applied Optics, Vol. 44, No. 9, 20 March 2005.
6. National Research Council, "Earth Science and Applications from Space: National Imperatives for the Next Decade and Beyond," Committee on Earth Science and Applications from Space: A Community Assessment and Strategy for the Future, The National Academies Press, ISBN: 0-309-10387-8, 2007.
7. F. Amzajerdian and M. J. Kavaya, "Development of solid state coherent lidars for global wind measurements" 9th Conference on Coherent Laser Radar, June 23-27, 1997, Linkoping, Sweden.
8. M. J. Kavaya, J. Yu, G. J. Koch, F. Amzajerdian, U. N. Singh, and G. D. Emmitt, "Requirements and Technology Advances for Global Wind Measurement with a Coherent Lidar: A Shrinking Gap," SPIE International Symposium on Optics & Photonics, Lidar Remote Sensing for Environmental Monitoring VIII, San Diego, CA, August 26-30, 2007.
9. R. R. Baize, F. Amzajerdian, R. Tolson, J. Davidson, R. W. Powell, and F, Peri, "Lidar Technology Role in Future Robotic and Manned Missions to Solar System Bodies," Symposium of Advanced Devices and Materials for Laser Remote Sensing, Materials Research Society Proceedings, Vol. 883, 2005.
10. A. E. Johnson, A. R. Klumpp, J. B. Collier, and A. A. Wolf, "Lidar-Based Hazard Avoidance for Safe Landing on Mars," AIAA Journal Of Guidance, Control, and Dynamics, Vol. 25, No. 6, 2002.
11. Wong, E.C., et al.,"Autonomous Guidance and Control Design for Hazard Avoidance and Safe Landing on Mars", AIAA Atmospheric Flight Mechanics Conference and Exhibit 5-8, 4619, Monterey, California, August 2002.

Mater. Res. Soc. Symp. Proc. Vol. 1076 © 2008 Materials Research Society   1076-K04-02

ALADIN Doppler Wind Lidar and Related Programs at EADS Astrium

Didier Morancais, Frédéric Fabre, and Yves Toulemont
EADS Astrium, 31 rue des Cosmonautes, Toulouse, 31402 Cedex, France

ABSTRACT

The Atmospheric Laser Doppler Instrument (ALADIN) is the payload of the ADM-AEOLUS mission, which will make direct measurements of global wind fields. It will determine the wind velocity component normal to the satellite velocity vector. The instrument is a direct detection Doppler Lidar operating in the UV, which will be the first of its kind in space.

ALADIN is now in its final construction stage: the integration of the Flight Model is on-going. Most of the subsystems have been integrated; the payload performance and qualification test campaign will commence.

This paper describes the ALADIN development status and the results obtained at this stage. This regards the receiver performance, the telescope development and the challenges of the laser.

The paper will also provide insights on the ATLID instrument design which is the backscatter lidar for the EarthCARE mission. This lidar program is starting its detailed design phase.

The ALADIN and ATLID instruments are developed by EADS Astrium Satellites for the European Space Agency.

AEOLUS MISSION

ALADIN (figure 1) is a Doppler Lidar operating in the near ultra-violet spectral region using backscatter signals from aerosol at low altitude and from air molecules at high altitudes. ALADIN will produce more than 3 thousand wind profiles per day, each one providing wind velocity from ground up to 30 km altitude: this is equivalent to the worldwide radiosonde system currently used by meteorological offices. The wind profile vertical resolution is 500 m at low altitude and 2 km at the highest altitudes. The wind measurement accuracy is better than 1m/s in the Planetary Boundary Layer and better than 2 m/s in the Troposphere. The wind profile data will be processed and delivered in less than 3 hours after the measurements.

The measurement principle is shown in figure 2. The satellite trajectory is a Low Earth Orbit at 400 km altitude with sub-satellite point motion around 7 km/s. The instrument points towards the Earth with a 35° slant angle versus the Nadir direction, oriented across-track. It measures the projection of the horizontal wind onto the inclined line-of-sight, for every altitude. In order to obtain high accuracy, 700 wind profiles are averaged during 7 seconds, which corresponds to a 50 km line on ground. This measurement is repeated every 28 seconds, which corresponds to every 200 km along the orbit.

Figure 1. Artist's view of AEOLUS satellite © ESA

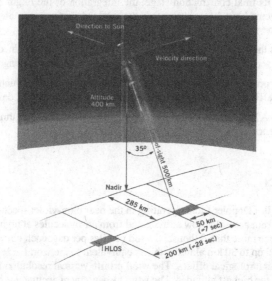

Figure 2. Measurement principle of the Atmospheric Laser Doppler Instrument (ALADIN)

ALADIN DESIGN OVERVIEW

The ALADIN instrument [1], [2] emits laser pulses towards the atmosphere and measures the Doppler shift of the return signal, backscattered from different altitudes in the atmosphere. The instrument emits the pulses at the 355 nm wavelength with a high power solid-state laser [3] featuring high efficiency and high reliability. Both the emitter and receiver use a large telescope. The receiver analyses the backscattered signal from the atmosphere: the receiver combines a fringe-imaging channel (analyzing low altitude aerosols and clouds) and a double-edge channel (analyzing air molecules). The two scattering mechanisms have different properties and allow coverage of the full range of altitudes. The instrument is designed to pave the way for potential future operational missions.

ALADIN is composed of two blocks: the instrument core which is located outside the platform and the remote electronics located inside. The instrument core includes the main structures and baffle, the telescope, the laser heads, the receiver optics and detectors, and two mechanisms (laser redundancy switch and Laser Chopper). The Core is fixed onto the Nadir side of the platform. The opto-mechanical architecture is based on a mono-static concept: the transmit and receive beams propagate through the same telescope. This architecture allows one to limit the field-of-view, hence to improve the daytime performance, and to relax the telescope and optics stability requirement. The optical protection of the receiver during laser emission is ensured by a chopper mechanism, located at the receiver entrance. The ALADIN instrument view is shown in figure 3.

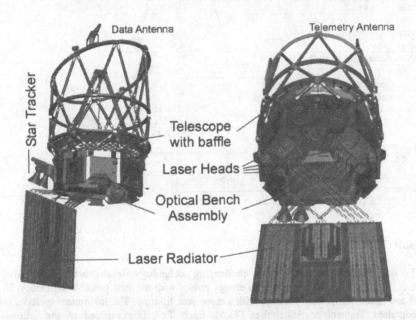

Figure 3. ALADIN Instrument View

The Transmitter is based on a Power Laser Head (PLH), seeded by a Reference Laser Head (RLH). The PLH and RLH are cooled by two radiators mounted on the anti-sun side of the platform connected via a set of heat pipes. The routing of the heat pipes is designed to allow on-ground testing of each PLH. Two sets of laser heads are implemented on the main structure (cold redundancy), the switching being performed by a FFM (Flip-Flop Mechanism).

The remote Electronics include the Detection Electronic Units (DEU), the Transmitter Laser Electronics (TLE) and the ALADIN control and data management units (ACDM).. The design includes a full redundancy for the remote electronics, with box-redundancy for the ACDM and TLE and internal redundancy for the DEU. All the electrical units are implemented inside the satellite bus. This provides thermal and mechanical decoupling from the instrument core. The DEU operates the Detector Front-end Units: timing sequence (main clocks), secondary power and video signal digitisation functions. The TLE operates the two laser heads and provides power supplies and control functions. The ACDM operates the DEU and TLE, and provides control (synchronisation) of all equipments, power and data buses to these units and interface with the platform.

Table 1. ALADIN main characteristics

Feature	Value
Operating wavelength	355 nm
Transmitted energy per pulse	120 mJ
Pulse repetition frequency	100 Hz
Telescope diameter	1.5 m
Total instrument field of view (full angle)	22 μrd
Transmitter frequency range	+/- 12.5 GHz
Rayleigh spectrometer frequency range	+/- 1.1 GHz

Rayleigh spectrometer characteristics :
    flter spacing
    filter FWHM
	
	5.5 GHz
	1.6 GHz

Mie spectrometer resolution	135 MHz
CCD quantum efficiency	80 %
Detection chain resolution	16 bits
Detection chain noise (50 accumulated shots)	4 e-/pixel rms
Mass	500 kg
Power consumption	840 W

TRANSMITTER LASER

The transmitter laser is the most challenging technology development of the Aladin instrument. The laser has to provide high energy pulses with the best possible efficiency, in a compact and space compatible package with a three-year lifetime. The instrument includes two fully redundant Transmitter Assemblies (TxA). Each TxA is composed of the following elements:
1. Power Laser Head (PLH)
2. Reference Laser Head (RLH), connected to the PLH via fiber optics
3. Transmitter Laser Electronics (TLE)

The PLH is a diode-pumped Q-switched Nd:YAG laser working in the third harmonic. This laser emits a single axial mode, obtained via injection seeding. The PLH produces the high energy laser pulse of 120 mJ at 355 nm with a pulse repetition frequency of 100 Hz. The pulses are emitted in "bursts" of 12 seconds, repeated every 28 seconds, which allows reduction of the demanded power.

The PLH is composed of the following modules: Injection Optics, Master Oscillator, two Amplifiers and the Harmonic Generation section. The Master Oscillator is built on an end-pumped Nd-YAG rod and the Amplifiers are design with side-pumped Nd-YAG slabs. Each Power Laser Head includes 18 laser diodes stacks, each capable to provide 1000 W optical peak power but used at derated power. The third harmonic is generated by two non-linear crystals. The PLH is conductively cooled internally via proper thermal design and externally via the instrument heat pipes

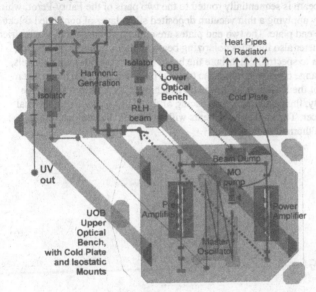

Figure 4. Power Laser Head Optical Layout

The RLH is a diode-pumped Nd:YAG double monolithic laser that produces a CW beam (injection seeder) for the PLH. This beam is fed into the PLH by means of an optical fibre. It is frequency-tunable in order to allow the receiver calibration and spectral registration.

RECEIVER

The ALADIN receiver performance capabilities are based on technology improvement in detection (CCD) and optics (very high resolution filters), developed and tested in earlier R&D programs, including several patents.

The optical bench assembly is the receiver function. It features a Transmit/Receive Optics (TRO) which includes a passive diplexer using a polarizing beam splitter, relay optics with two foci (one chopper, one field stop), and a small bandwidth filter to limit the Earth radiometric background. The chopper mechanism is closed during laser firing in order to block straylight going into the receiver. "Aberration generators" are included in the TRO emitting and calibration paths to adapt the laser divergence to the eye safety criteria.

The aerosols ("Mie") receiver is based on a Fizeau interferometer. The incoming beam from the Rayleigh receiver is linearly polarized. The Mie receiver integrates a beam expander to adapt the input beam diameter to the size of the Fizeau interferometer. A quarter-wave plate rotates the reflected beam polarization by 90° back to the Rayleigh receiver. The fringes generated at the output of the Fizeau interferometer are imaged on the detection unit with an afocal lens.

The molecules ("Rayleigh") receiver is based on a double edge sequential Fabry-Perot interferometer. The input beam is sequentially routed to the two parts of the Fabry-Perot, which slightly differ in spacing by applying a thin vacuum deposited silica layer of controlled thickness on the inner surface of one end plate. The two end plates are optically contacted to a cylindrical silica spacer. The spectrometer also includes polarizing beam splitters, quarter wave plates, mirrors and focusing optics to spectrally distribute the beam between the different channels, taking advantage of the change of orientation of the circular polarization of the beam when reflected. Spectral tuning of the Rayleigh spectrometer is achieved by slightly varying the temperature of the assembly, thus expanding the spacing of the plates according to thermal expansion of the silica spacer. To avoid any gradients within the unit, the temperature is radiatively controlled via a thermal enclosure.

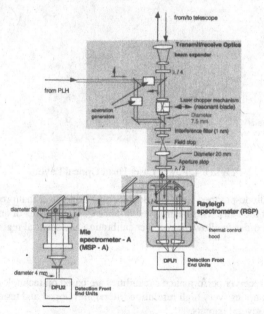

Figure 5. Receiver optical architecture

The reference detector for both Rayleigh and Mie receivers is the Accumulation CCD, a small area CCD with a 16x16 useful image zone and two interleaved memory zones (Astrium patent). This geometry and a specific sequencing of charge transfer allow on-chip accumulation of laser shots without mixing the vertical samples. The vertical wind profile is built-up in the memory zone and read-out at low frequency after accumulation; hence the read-out noise occurs only one time for the cumulated shots. This concept adds the benefits from quasi photon counting mode operation and high quantum efficiency as provided by thinned and back-illuminated devices. This detector provides quasi-perfect detector performance, high dynamic range, operation versatility, and makes the detection chain electronics design and operation simple and robust.

TELESCOPE

The instrument uses a very lightweight telescope made of silicon carbide: this technology allows one to select a large diameter while keeping a low mass and high stiffness; hence improving the overall performance. The Silicon Carbide (SiC) material offers a significant mass saving advantage over conventional technologies, especially when the complete telescope, mirror and tripod structure, is made with the same material.

The Aladin telescope, developed by EADS Astrium, is a 1.5 m diameter afocal Cassegrain telescope, with low mass (75 kg) and high first frequency (60Hz). The focus is thermally adjusted by means of heaters. This design avoids implementing a refocusing mechanism. The secondary mirror is fixed via a tripod structure. The telescope is fixed on the main structure via titanium. Telescope fixations are located at two thirds of the primary mirror diameter. The architecture allows a direct transfer of telescope loads to the platform interface and also optimizes both the load spreading in the M1 and its stiffness.

The driving parameters for the telescope design are the focus-fit (overall inter-mirror M1-M2 distance) that shall be maintained within a few microns. Due to the use of the same material, the focus is not sensitive to a uniform temperature variation. In addition, the high conductivity of the SiC material avoids thermal gradients detrimental to the stability and optical performances.

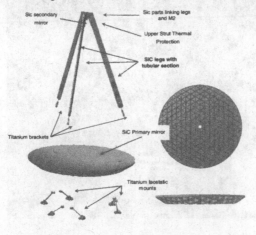

Figure 6. Telescope and Primary Mirror design

ALADIN DEVELOPMENT STATUS

The model philosophy aims at reducing the development risk while maintaining a short planning duration compared to other instruments with similar complexity. Two instrument models are developed: an Optical-Structural-Thermal Model (OSTM) and a Flight Model (FM).

The OSTM allowed to perform the mechanical and thermal qualification of the Core Instrument. A Shack-Hartman configuration was used to measure the telescope wavefront error, defocus and line-of-sight stability. The OSTM has been submitted to thermal vacuum and optical tests. The OSTM was then mounted onto the satellite and submitted to mechanical tests (sine, acoustic and shock).

All the FM equipments have been delivered and the transmitter laser is in the final test phase. The receiver equipments have been integrated and the receiver assembly performances have been verified, allowing to consolidate the instrument performance. The receiver is now being aligned onto the instrument core.

The Telescope has been fully integrated and aligned. The wavefront error has been verified to be lower than one 350 nm rms and the focus was adjusted within +/- 2 µm accuracy.

The Transmitter equipments have been completed and the assembly level tests have started. All the functional and optical performances have been verified on the FM1 assembly at ambient conditions. The mechanical qualification has been performed on the Engineering Model and will be repeated on the two flight models. The thermal vacuum tests of the FM1 assembly have been delayed to a defect component which has now been replaced.

The main challenges related to the transmitter are the laser diode qualifications, coatings qualification (especially damage threshold vs number of shots). These issues are now solved thanks to extensive qualification campaigns on those components.

The ALADIN instrument integration is on-going with a delivery planned in 2009.

Figure 7. ALADIN Flight Telescope

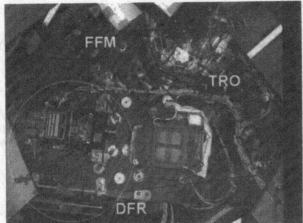

Figure 8. ALADIN Receiver

Figure 9. ALADIN Power Laser Head

ATLID INSTRUMENT OVERVIEW

The ATLID instrument is a new European lidar program, part of the EarthCARE mission. This satellite is dedicated to the study of Earth radiative transfer and related climate models. The two main instruments are a cloud radar and the backscatter lidar. ATLID is dedicated to the measurement of the aerosols concentration and the cloud physical parameters.

The instrument is based on a tripled-frequency Nd-YAG laser associated to a high resolution receiver in order to discriminate between aerosols and molecular backscatter. The laser features however medium energy at 355 nm (30 mJ/70 Hz) and is frequency tuned. The receiver features a very low noise detection chain based on a CCD, with a principle adapted for high vertical resolution. The telescope is also made of Silicon Carbide having a diameter of 60 cm. The instrument budgets are 300 kg and 500 W. The heritage acquired from ALADIN is significantly re-used. In addition, new technologies developed by Astrium are included such as Loop Heat Pipes, which provide accommodation and testing flexibility with regards to conventional heat pipes.

The program has started its detailed design phase in early 2008 and the instrument delivery is planned in 2012.

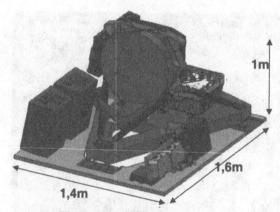

Figure 10. ATLID overview

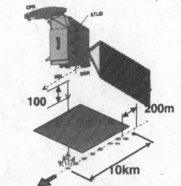

Figure 11. ATLID measurement principle

CONCLUSION

The ALADIN instrument will be the first European lidar and the first Doppler Wind Lidar in space. This challenging programme has progressed well with satisfactory results as the completion of the OSTM program, the completion of the integration and tests of the Flight Receiver and Telescope and the successful Laser components qualification.

The major activities that will happen in the near future are the Transmitter assembly qualification, the instrument final integration and the final qualification.

The detailed design phase of the ATLID instrument has started this year. This payload will be the second European lidar in space and a backscatter lidar.

EADS Astrium Satellites are developing the two instruments for the European Space Agency and is the leader of active optical instruments in Europe.

REFERENCES

1 "Atmospheric Dynamics Mission, Report for Mission Selection", ESA-SP-1233(4) (1999)

2. D. Morançais et. al.: "ALADIN, the first wind lidar in space : development status", *International Conference on Space Optics* (2006)

3. A. Cosentino et al., *International Conference on Space Optics*,(2006)

Mater. Res. Soc. Symp. Proc. Vol. 1076 © 2008 Materials Research Society   1076-K04-03

Geospatial Remote Sensing Using Advanced Sensor Systems

A. Vaseashta

Graduate Program in Physical Sciences, on detail from Nanomaterials Processing and
Characterization Labs., One John Marshall Drive, Huntington, WV, 25575

ABSTRACT

Increased demand on monitoring, surveillance, and communication has necessitated
satellites with high resolution, accuracy, speed, and authenticity. Phenomenon of plasmonic
interactions in nanomaterials has further expanded the realms of possibilities available so far
with conventional opto-electronic devices. The nanophotonics structures coupled with
lightweight structures and advanced nanotechnology based sensors have resulted in launching of
nanosatellites by several countries. The use of nanophotonics in conjunction with integrated
micro/nano optoelectronic technologies in space will reduce susceptibility of the system to EMI,
weight/volume of cables, and propagation loss while enhancing signal processing speed, spatial
resolution near-field imaging, information transmission and storage capacity, and security
encryption capabilities. Use of nanomaterials based advanced sensor systems in satellites and
aerial remote sensing science and methodologies to improve performance, resolutions, and
security is investigated. Use of improvised satellite systems in low to medium earth orbit to
achieve medium to high resolution, wide swaths and low noise equivalent reflectance is
envisaged.

INTRODUCTION: POLLUTION, ENERGY, AND SECURITY

Anthropogenic pollution, energy, and international security rank among the top three
concerns facing the 21st century. The U.S. Census Bureau reports the world population is about
6.8 B people as of May 2008 with projections that reach 10 B by 2010. The ever increasing
demand on energy and resulting increase in pollution has reached alarming levels in some
developing countries. It is widely known that long-term exposure to air pollution provokes
inflammation, accelerates atherosclerosis, and alters cardiac function. Within the general
population, medical studies suggest that inhaling particulate matter (PM) is associated with
increased mortality rates which are further magnified for people suffering from diabetes, chronic
pulmonary diseases, and inflammatory diseases. Gaseous air pollutants, like NO_x, SO_2, CO, and
CH_4, are some of the primary air pollutants in urban and industrial areas. Secondary pollutants
are created from the primary pollutants by complex photochemical reactions in the presence of
ultra-violet (UV) radiation forming free radicals. Sulfur and oxides of nitrogen (NO_x) from
industrial emissions transforms into ammonium sulfate and nitrate. In the presence of
atmospheric moisture, NO_x transforms into HNO_3 and HNO_2. The role of HNO_3 is central in
various master physio-chemical processes occurring in earth's atmosphere as a reservoir
molecule of NO_x species form "smog". On a related note, the global war on terrorism calls for
increased demand on monitoring and surveillance. Furthermore, accurate weather prediction and
environmental pollution detection and monitoring call for geospatial sensing with high
resolution, accuracy, speed, and authenticity.

Satellites play a major role in communication, navigation, climatology, surveillance, and environmental monitoring and support a multitude of applications, which range from air traffic control services and support functions in managing navigational data and digital communications; broadband digital communications for transmitting high-speed data and video directly to consumers and businesses; direct-broadcast satellites for direct reception by the general public; to environmental monitoring including the vertical thermal structure of the atmosphere, the movement and formation of clouds, ocean temperatures, snow levels, glacial movement, and volcanic activity. Satellites providing fixed-satellite services (FSS) transmit radio communications between ground Earth stations at fixed locations to broadcast media events. X-band satellite services to governments is a new application using specially allocated frequency bands and waveforms and are used to gather intelligence. Over last couple of decades, satellite performance has substantially increased in terms of higher information transmission capacity, opto-isolation of critical spacecraft subsystems, high speed optical processing of RF and microwave signals, low propagation loss, enhanced security encryption capabilities, higher resolution, wider swaths and low noise equivalent reflectance. Materials approaching nanoscale dimensions exhibit characteristics with numerous unique and hitherto unexploited applications. Nanomaterials are a fundamentally and entirely new class of materials with remarkable electrical, optical, and mechanical properties. Advances in synthesis and characterization methods have provided the means to study, understand, control, or even manipulate the transitional characteristics between isolated atoms and molecules, and bulk materials [1]. Nanotechnology and micro/nano electromechanical systems (MEMS/NEMS) hold the potential to revolutionize the field of satellite design.

Hence, efforts are underway to use nanomaterials in satellite and aerial remote sensing science and methodologies in combination with integrated mico/nano optoelectronic technologies and MEMS/NEMS for reduced susceptibility of the system to EMI, reduced weight of the signal cables (< 1/30), higher information transmission capacity (GHz), reduced weight and volume, opto-isolation of critical spacecraft subsystems, high speed optical processing of RF and microwave signals, low propagation loss, and enhanced security encryption capabilities. Carbon Nanotubes (CNTs) based field emission electron guns (FEG) employ low voltage for emission and are actively researched as cold cathode microwave generation devices [2]. CNT based composites provide light-weight and compact platforms for compact design with mechanical and thermal robustness. Such satellite systems can be placed in low Earth orbit (LEO) to medium Earth orbit (MEO) with multi-sensor satellite imagers, to achieve medium to high resolution, wide swaths using negative refractive index surfaces and noise equivalent reflectance (NER) values of less than 0.5%. The resulting "nanosatellites" will contain all of the typical subsystems and use distributed satellite systems defined as constellations or clusters of satellites working in concert to achieve missions; e.g. global positioning and navigation, communication and information transfer, reconnaissance, space science, security surveillance, and environmental monitoring and sensing.

SATELLITES IN REMOTE SENSING

Remote sensing using satellites started with the launch of Landsat MSS-I (Multi-Spectral Scanner) in 1972. Subsequent Landsat MSS series and Landsat Thematic Mapper (TM) series of satellites revolutionized earth observation from space. Satellite image data has traditionally been unexploited for atmospheric pollution studies. Observation of atmospheric parameters using satellite has recently become important due to risk of increasing pollution and possible linkage with climate change. Satellite image data consists of earth radiances observed by its sensors in different bands. For thermal infra-red (TIR) bands, the radiances represent a function of the temperature, emissivity of the ground surface and the atmospheric column above and its surroundings. Satellite image data can aid in detection, tracking, and understanding of pollutant sources and transport by providing observations over large spatial domains, with three dimensional models (3-D Models). Satellite data can be used quantitatively to validate air quality models. The pollution assessment of optical atmospheric effects can be quantified in terms of aerosol optical thickness (AOT). With the launch of NASA's Terra satellite system, a part of the Earth Observing System (EOS) in December 1999; satellite observation of atmospheric parameters are easier to acquire. Using the Differential Optical Absorption Spectroscopy-Method (DOAS) algorithm, University of Bremen scientists have analyzed the data of the Global Ozone Monitoring Experiment (GOME) on ERS-2 and the SCIAMACHY on ENVISAT to retrieve NO_2 columns with high accuracy in the 425-450 nm regions. The data can be modelled for any specific city and correlated with data from ground based stations, such as the U.S. environmental protection agency (EPA), Texas Commission on Environmental Quality (TCEQ), France's AIRPARIF, the Netherlands' National Institute for Public Health and Environment (RIVM), British Atmospheric Data Centre (BADC), and the Natural Environmental Research Council (NERC), to name a few.

There is observational evidence that these aerosols can alter cloud properties. Scattering albedo increases with pollution, thus decreasing backscatter fraction. Aerosol radiative forcing depends on hygroscopicity, which in turn depends on aerosol photo-chemistry. IR absorption spectroscopy has played an important role in the identification of trace pollutants in both ambient air and synthetic smog systems. Employing Advanced Space borne Thermal Emission and Reflection Radiometer (ASTER) data, we have investigated air pollutants using satellite images [3] using high spatial and spectral resolution images from the TIR and SWIR bands and compared the satellite data with the EPA air quality and spectroscopic study data of atmospheric pollution. A conceptual model of the interplay of data from satellite and ground based stations is shown in figure 1 [4]. ASTER is an imaging instrument flying on the Terra satellite as part of NASA's Earth Observing System (EOS). It is the only high-spatial resolution instrument on the satellite that has 14 bands from the visible to the thermal infrared region. In the visible green and near-infrared (V-NIR) range between $0.52 - 0.86 \,\mu m$, there are three bands with 15 m resolution. In the SWIR range between $1.6 - 2.43 \,\mu m$, there are six bands with 30 m resolution. In the third range between $8.125 - 11.65 \,\mu m$, there are five bands with a resolution of 90 m. ASTER acquires data over a 60 km swath of which center is pointable.

The data as observed from satellite is processed employing the Multivariate techniques, which are based on groupings in a multivariate data set. Remote sensing software ER- Mapper™ (v.7.x) is generally employed to accurately model the data through feature extraction processes for pattern recognition. Several digital imaging processes such as geometric image registration, radiometric normalization, principal component analysis, and data fusion are employed to process the image for accurate feature extraction. Multidate correlations and regression analysis methods are used for individual gases with satellite-recorded reflectance of bands.

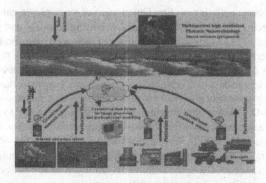

Figure 1: Conceptual model showing pollution monitoring using Satellite and ground Stations [2].

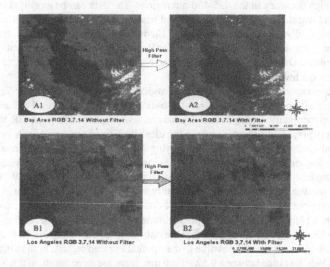

Figure 2: (a1): San Francisco data in 3,7,14 bands. (a2): San Francisco in RGB 3,7,14 with filter. (b1): Los Angeles area in RGB 3,7,14. (b2): Los Angeles area in RGB 3,7,14 with filter [4].

Although careful examination of images in figure 2 can model smog, particulate matter, and specific gaseous columns; the imaging technology is based on charge coupled devices (CCDs) and silicon photodetectors thus posing serious limitations in terms of resolution. As shown in Figure 3, VNIR, SWIR, and ThIR regions are limited to a very narrow range. To enhance sensitivity, resolution, and speed; it is critical that the latest advances in nanotechnology should be explored to replace the satellite components.

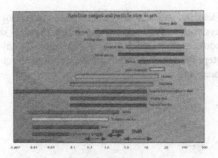

Figure 3: Satellite ranges and particles size in μm.

NANOTECHNOLOGY IN REMORE SENSING SATELLITES

Mie's theory represents two theoretical models, viz.: the classical electrodynamics for the propagation of light and the solid (or liquid) state theory for the interaction of light with the particle expressed by the complex, frequency dependent dielectric function, DF, of the particle material, which can mathematically be described as:

$$\varepsilon(\omega) = \varepsilon_1 + i\ \varepsilon_2 = \varepsilon_{Drude} + \chi_{interband}$$
$$= 1 - (n\ e^2/\ \varepsilon_0\ m_e)/(\omega^2 + i\ \gamma\ \omega) + \chi_{interband}$$

with n, e, ε_0, m_e, γ are respectively the electron density, elementary charge, field constant, effective mass and relaxation frequency of the conduction electrons of the Drude-Lorentz-Sommerfeld theory, $\gamma = 1/\tau$ with τ the Drude relaxation time and $(n\ e^2/\ \varepsilon_0\ m_e)^{1/2} = \omega_p$ the "Drude plasma frequency". The interband transitions exist in most metals thus causing nanomaterials to demonstrate hybrid surface plasmon polariton (SPP) excitations. They consist of collective electronic excitations in the conduction band and (complex) polarization including deeper lying bands. Most nanomaterials exhibit optical absorption and scattering spectra with complex multi-peak features, characterized by quantities viz., peak height, spectral peak position ω_{Max} and band width Γ. The DF includes electronic size and surface/interface effects, present in a nanoparticle. Although a detailed description is described elsewhere, it suffices to state that Mie's theory is used analogously to Fresnels formulae which was derived for samples with planar geometry. Extension of the theory yields effective permeability and refractive index to produce negative refractive index materials (NRIM) by plasmonic materials to image sub-λ features for producing near field super lens [5,6], enhancing swath of satellites. The inverted Mie theory yields the realistic DF including all electronic and optical size and surface/interface effects. The resulting evanescent field strongly interacts with adsorbed molecules, thus influencing the spectra. The application of inverse Mie theory is essential for structural, electronic and optical effects for selective detection substances in aqueous/bio-surroundings [7] using plasmonic nanostructures. Numerous associated applications of nanomaterials in satellites encompass quantum wire interconnect [8], high power electrochemical capacitors [9], data storage devices [10], THz oscillators for communication [11], low-voltage field emitters for X-ray generation [2], power sources using betavoltaics, and use of CNTs in light weight and heat-dissipating composites

25

[11]. Integrating plasmonic elements with existing Silicon based technology will likely result in plasmonic nanophotonic arrays with multifunctional capability. Figure 4 shows subsystems which are considered for nanomaterials based devices and systems. To advance the security goals of controlled access, integrity, and confidentiality we developed security risk analyses to identify threats and countermeasures.

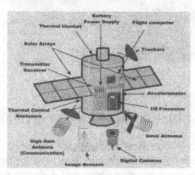

Figure 4: Nanosatellites with Nano Based components

CONCLUSIONS AND FUTURE DIRECTIONS

The field of satellite design continues to evolve due to complex missions, higher performance requirements, and evolving technologies. While satellites subsystems are designed for a specific mission, the trends toward modular design will likely to continue. Use of nanophotonics and combination of micro, nano, integrated opto-electronic technologies are envisioned to provide light weight, high performance, and cluster satellite configuration for monitoring specific missions.

Reference

1. A. Vaseashta, and I. Mihailescu, (editors) - Nanoscale Materials, Devices, and Systems for Chem.-Bio Sensors, Photonics, and Energy Generation and Storage. Springer (2008).
2. A. Vaseashta, J. Materials Science. 14, 653 (2003).
3. A. Vaseashta, et al., Science and Technology of Advanced Materials, 8(1-2), 47-59 (2007).
4. P. Roy, J.O. Brumfield and A. Vaseashta, Nanotech 2007, Technical Proceedings of the 2007 NSTI Nanotechnology Conference and Trade Show, 2, 631-634 (2007).
5. V. P. Drachev, V. Nashine, M. D. Thoreson, D. Ben-Amotz, V. J. Davisson, and V. M. Shalaev, Langmuir 21(18), 8368-8373 (2005).
6. J. B. Pendry, Phys. Rev. Lett. 85(18), 3966 (2000).
7. A. Vaseashta and J. I. Irudayaraj, J. of Optoelectronics and Adv. Mat. 7(1), 35-42 (2005).
8. L. Rotkina, J. F. Lin and J. P. Bird, Appl. Phys. Lett. 83(21), 4426 (2003).
9. Ch. Emmenegger et al., J. of Power Source, 124, 321 (2003).
10. M. S. Fuhrer et al., Nano Lett. 2(7), 755 (2002).
11. J. Yoon et al., Proceedings IEEE - ICMENS, (2003).
12. A. Vaseashta, Appl. Phys. Letters, 90(9), 093115 (2007).

Mater. Res. Soc. Symp. Proc. Vol. 1076 © 2008 Materials Research Society   1076-K04-04

Development Effort of the Lunar Orbiter Laser Altimeter Laser Transmitter

George B Shaw, Anthony W Yu, and Anne-Marie D Novo-Gradac
Laser & Electro-Optics Branch, NASA Goddard Space Flight Center, Mail Code 554, Greenbelt, MD, 20771

ABSTRACT

The Lunar Orbiter Laser Altimeter (LOLA) is one of six instruments on the Lunar Reconnaissance Orbiter (LRO) with the objectives to determine the global topography of the lunar surface at high resolution, measure landing site slopes and search for polar ices in shadowed regions. The LOLA laser transmitter is a passively Q-switched crossed-Porro resonator. The flight laser beryllium bench houses two oscillators (a primary oscillator and a cold spare). The two oscillators are designed to operate sequentially during the mission. The secondary laser will be turned on if the primary laser can no longer provide adequate scientific data products. All components used in the laser have space flight heritage. In this paper we will summarize the development effort of the LOLA laser including the material choice, design criteria and contamination control as applied to the flight laser build.

INTRODUCTION

Announced in 1961 and ended in 1975, NASA's Apollo program successfully conducted six manned landings on the Moon between 1969 and 1972. In early 2004, NASA defined its new Vision for Space Exploration whose overarching strategy is to conduct investigations and prepare for future human exploration of the Moon, Mars and beyond. A manned mission to the Moon is slated to occur by 2020, but could happen as early as 2015. NASA's return to the moon is initiated with the Lunar Reconnaissance Orbiter (LRO) [1,2]. The spacecraft is scheduled to be launched from the Kennedy Space Center in late 2008 and will spend at least a year mapping the surface of the Moon. Data from the six instruments aboard the orbiter will be evaluated to help select safe landing sites for future manned missions, identify prospective lunar resources and study the Moon's radiation environment. The LRO (figure 1(a)) baseline mission is one Earth year at 50 ± 20 km circular, polar orbit.

Figure 1. (a) Depiction of LRO showing the six scientific instruments as well as the Mini-RF technology demonstration; (b) Rendering of the LRO spacecraft in lunar orbit.

Lunar Orbiter Laser Altimeter

The Lunar Orbiter Laser Altimeter (LOLA) is one of seven instruments on the LRO spacecraft. The LOLA mission objectives are to produce a high resolution three-dimensional map of the Moon's surface, measure slopes and surface roughness of potential landing sites, and search for the presence of water ice in permanently shadowed regions. The LOLA instrument shown in figure 2 (a) pulses a single laser through a diffractive optical element (DOE) [3] to produce five beams that illuminate the lunar surface. For each beam, LOLA measures time of flight (range), pulse spreading (surface roughness), and transmit/return energy (surface reflectance). With its two-dimensional spot pattern (figure 2 (b)), LOLA unambiguously determines slopes along and across the orbit track. Analysis of the data at cross-over points from multiple orbital tracks will also provide insight on the gravitational field of the moon and its center of mass.

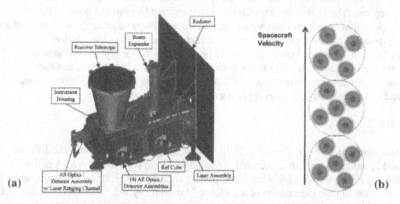

Figure 2. (a) Depiction of the LOLA Instrument. The instrument housing, laser bench, beam expander tube, receiver telescope tube, detector housings, laser electronics frame and radiator are all machined from I220H, Grade 1 beryllium. (b) Five spot beam pattern on the lunar surface. The beams are clocked ~ 26° to provide five adjacent profiles, 10 to 12 meters apart over a 50 to 60 meter wide swath, with combined measurements in the along track direction every 10 to 12 meters [3]. The red circles illustrate the laser spots on the ground (100 μrad) and the gray circles portray the receiver field of view (400 μrad) [3].

LOLA FLIGHT LASER TRANSMITTER

The LOLA laser transmitter follows the continued success of the Geoscience Laser Altimeter System (GLAS) [4] and MESSENGER Laser Altimeter (MLA) [5] in using a diode pumped, Cr:Nd:YAG slab with passive Q-switch and a crossed-Porro resonator configuration. The laser transmitter consists of an I220H, Grade 1 beryllium (Be) flight laser bench housing two oscillators (a primary oscillator and a cold spare), an I220H, Grade 1 Be 18X transmit beam expander and a DOE. The two output beams are polarization combined along the beam combining stage as shown in figure 3(a). The two lasers are designed to operate sequentially during the mission with the secondary laser turning on if the primary laser can no longer provide adequate scientific data products. A picture of the completely assembled LOLA laser on Be bench is shown on the next page in figure 3(a). Figure 3(b) shows the fully assembled LOLA Instrument consisting of the Be laser bench integrated in the Be instrument housing along with the laser electronics assembly (LEA) receive detectors, transmit beam expander and receive telescope, and radiator. The instrument has completed final ambient testing and is awaiting thermal blanketing before undergoing thermal vacuum (TVAC) testing.

Figure 3. (a) Fully assembled LOLA Flight Model Laser (dimensions: 5" x 7" x 1.5"). (b) Fully integrated LOLA Flight Instrument, mounted on TVAC fixture awaiting thermal blanketing.

LOLA LASER DESIGN CONSIDERATIONS

Mechanical

Very early on in the LOLA Laser design phase Be was chosen as the material for the laser bench due to its attractive mechanical and thermal properties. Beryllium is the lightest known structural metal ($\rho = 1.85$ g/cm^3) yet it is very stiff with Young's modulus of 287 GPa [6]. By comparison, the density of Be is almost 25% smaller than aluminum (Al) ($\rho = 2.70$ g/cm^3) and the Young's modulus for Be is more than four times that of Al (70 GPa) [6]. Despite these desirable mechanical properties, they do not come with significant thermal disadvantages. The coefficient of thermal expansion (CTE) of Be is 12 μm/(m·K), nearly 50% less than the CTE for Al (23.5 μm/(m·K) [6]. Although the other properties compare favorably, the thermal conductivity of Be (201 W/m·K) is slightly less than Al (237 W/m·K) [6], but it is still very good. Finally, the specific heat of Be (1.82 J/(g·K)) is more than twice that of Al (0.90 J/(g·K)) and nearly half the specific heat of water (4.12 J/(g·K)) [6].

Clearly, lightness is an important consideration for aerospace applications. But for a space laser whose alignment must remain precise, without adjustment, over a wide thermal range, the other properties of Be are of more concern. Because Be is very stiff and has a low

CTE, it is an ideal choice for space laser applications. Furthermore, the high specific heat shows that Be is not going to react too quickly to rapid changes in the exterior thermal environment. These qualities will help the LOLA Laser to remain stable in lunar orbits that will take the spacecraft from direct sunlight on one side of the moon to total eclipse on the dark side during a given two hour orbit.

Optical

Because the LOLA Laser Transmitter will operate outside of the Earth's protective atmosphere and magnetic field, radiation in the form of Gamma rays and high energy protons are important concerns. These types of radiation can damage some transparent crystal media and cause them to darken. If this happens in the laser cavity, this damage could start to absorb the emission and lower the efficiency of the system to the point where the laser ceases to operate. Consequently, all of the optics in the laser cavities as well as the output stage need to be made from materials that are radiation hard. As mentioned in the previous section, the LOLA Laser is a crossed-Porro resonator that is based on a diode pumped Cr:Nd:YAG slab. The flight heritage for co-doping YAG with chromium (Cr^{3+}) as well as the absorbing neodymium ion (Nd^{3+}) stems back to the GLAS and MLA lasers. The presence of the Cr^{3+} ion in the Cr:Nd:YAG crystal matrix does not affect the gain quality of the medium, but it has been shown to reduce optical loss at the 1064 nm emission wavelength due to radiation-induced damage [7]. For similar reasons, radiation hard fused silica was used for the Porro prisms, rilsey wedges and tilt plates throughout the laser cavities and output stage. Certain other optics were unavailable as fused silica so the zero order wave plates were made from crystalline quartz and the polarizers were made from BK7/G18 glass. All of the laser optics have space flight heritage from the GLAS and MLA lasers.

In addition to the materials choices for the laser optics, we ordered large lots of each type of optic. We employed a rigorous screening program to assess the quality of our optics. Some of the criteria were performance related, *i.e.* how close was a given optic to our design specifications of what we ordered. Other criteria were based on detailed microscope inspections of the quality of the optic and its coating.

Contamination Control

To varying extents, all lasers are susceptible to contamination induced aberrations and failures. But in the laboratory, it is straightforward to diagnose the situation and replace any component that has failed for whatever reason. For space lasers, there are no opportunities to make changes or adjustments once it launches, so great care must be taken during the parts processing as well as the build phase to minimize contamination not only on a particle level but also on a molecular level. Every single piece part in the LOLA Laser, from the bench itself to the optical mounts and even the screws and washers, was precision cleaned with a specific solvent recipe and then the parts were rinsed with solvent. The rinsate was analyzed and the resultant non-volatile residue (NVR) per surface area of the part had to be less than 0.33 $\mu g/cm^2$. Furthermore, the constituents of the NVR could not contain any compounds that absorb light in the 1064 nm region (*e.g.* methyl silicones).

Like the metallic parts listed above, certain soft, porous parts such as plastics, electrical harness wiring and connectors, viton o-rings, etc., were cleaned and verified by the protocol listed above. But after verification, they were subjected to a vacuum bake out that sought to drive out any compounds that could be buried in the bulk material. This is of particular concern for the LOLA Laser because it is designed to operate in vacuum. There is a particulate filter on the instrument housing that allows the laser cavity volume to vent to the vacuum of space during the trip to the moon. Over time, undesired chemicals could outgas from these soft parts and contaminate the laser optics. Depending on the nature of the compound and the amount, it could lead to a film being generated on an optic that results in optical damage or degraded laser performance.

During the laser build phase for LOLA, full gowns were worn and work was performed in a clean room that exceeded Class 10000 (and approached Class 1000). In a zero or reduced gravity environment, any particles in the laser cavity are free to drift around and are likely to be attracted to the strong electric field of the laser beam. If particles settle on an optic in the laser cavity, this can lead to laser induced damage of the optic and degrade laser performance. As shown in figure 4, regular black light as well as white light inspections were conducted of the LOLA Laser, LOLA instrument housing, the surrounding work area and tools.

Figure 4. Black light and white light inspections and particle removal of the laser cavity surfaces just prior to final integration of the LOLA Laser into the LOLA Instrument housing.

Mission Lifetime

Ultimately, the LOLA Laser Transmitter has a mission requirement of one billion shots on orbit. The design decisions for the mechanical and optical components as well as the contamination control measures will help to ensure reliability for this laser. In addition, a design change was implemented from the Engineering Model to the Flight Model that permitted removal of one wave plate from the laser cavity. The benefit of this resulted in lowering of the intracavity fluence by approximately 50%, which should enhance the longevity of the LOLA Laser Transmitter.

AS DELIVERED PERFORMANCE NOTES

The as delivered performance of both primary and secondary lasers is summarized in figure 5 (next page). The far field patterns of both lasers after the transmit beam expander and DOE are shown in figure 6 (next page). The laser is designed to produce a 2.7±0.3 mJ per pulse (prior to transmit beam expander and DOE) at a rate of 28 Hz, pulse width of 6±2 ns. The far field divergence of each of the five spots is 100 μrad. At 50 km orbit, this will produce five-5 m diameter spots with center-to-center separation of 25 m on the lunar surface. The LOLA laser transmitter has successfully undergone vibration testing and temperature cycling in air. Currently the laser transmitter is awaiting instrument level thermal vacuum (TVAC).

Parameters		Requirements	Final Delviery Measured Data	
Date	Units		9/29/07	9/29/07
Chiller Temp	□C		18.00	
Laser S/N			L1	L2
Reference Temp	□C		19.43	19.43
Laser Diode Array Temp	□C		16.11	23.12
Pump Current	A	60 < I < 90	60.93	67.87
Rep Rate	Hz	28.0 +/- 0.1	28.00	28.00
Prime Voltage	V		27.94	27.93
Prime Current	A		0.432 to 0.453	0.464 to 0.480
Average Energy	mJ	2.7 +/- 0.3	2.63	2.97
Pulse switched out time	μs	130 < t < 210	153.20	161.80
Pulse Width	ns	6 +/- 2	4.92	4.98
Wavelength	nm	1064.3 +/- 0.1	1064.38	1064.41
Divergence - Center Spot	μrad	100	99.50	102.00
Divergence - Spot 1	μrad	100	103.00	104.00
Divergence - Spot 2	μrad	100	98.00	99.50
Divergence - Spot 3	μrad	100	100.50	104.50
Divergence - Spot 4	μrad	100	97.00	98.50

Figure 5. Performance requirements and final, as delivered performance results of the LOLA Laser Transmitter.

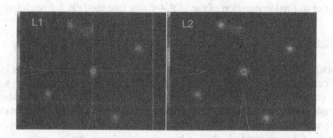

Figure 6. Primary (L1) and secondary (L2) laser far field patterns after the beam expander and DOE (the dimmer spot in the upper center is the return from the LOLA instrument alignment reference cube).

CONCLUSION

The LOLA instrument is designed to map the lunar surface and provide unprecedented data products in anticipation of future manned flight missions. The laser transmitter design was realized and verified with the engineering model (EM) laser. The LOLA laser transmitter design has heritage dated back to the MOLA laser transmitter launched more than 10 years ago and incorporates lessons learned from previous missions (GLAS, MLA). The most notable design change in the LOLA laser transmitter from previously flown missions is the use of two vendors laser diode arrays (LDA) with more than 40% derating. This change will help to enable the LOLA laser to fulfill the mission goal of 1-year continuous operation (or 1 billion shots) in lunar orbit.

ACKNOWLEDGEMENT

The authors acknowledge the support of the LOLA science team, especially David E. Smith (PI), Maria Zuber (PI), the LOLA instrument team, Glenn Jackson (instrument manager), Ron Zellar (deputy instrument manager), John Cavanaugh (system engineer) and Larry Ramsey (contamination lead).

REFERENCES

1. Chin, G. *et al.*, Space Sci. Rev. **129**, 391-419 (2007).
2. http://lunar.gsfc.nasa.gov/
3. Smith, J.G., *et al.*, "Diffractive Optics for Moon Topography Mapping," Proc. SPIE, Vol. **6223**, (2006).
4. Afzal, R.S. *et al.*, IEEE J. Sel. Top. Quant. Elect., **13**, 511-536 (2007).
5. Krebs, D.J., *et al.*, Appl. Opt. **44**, 1715-1718 (2005).
6. *CRC Handbook of Chemistry and Physics*, Ed. D. R. Lide, The Chemical Rubber Co., 1999.
7. Rose, T.S., Hopkins, M.S. and Fields, R.A., IEEE J. Quant. Elect. **31** (9), 1593-1602 (1995).

Mater. Res. Soc. Symp. Proc. Vol. 1076 © 2008 Materials Research Society 1076-K04-05

All Fiber Coherent Doppler LIDAR for Wind Sensing

Toshiyuki Ando[1], Shumpei Kameyama[1], Kimio Asaka[2], Yoshihito Hirano[1], Hisamichi Tanaka[3], and Hamaki Inokuchi[3]

[1]Information Technology R&D Center, Mitsubishi Electric Corporation, 5-1-1 Ofuna, Kamakura, 247-8501, Japan
[2]Communication Systems Center, Mitsubishi Electric Corporation, 8-1-1 Tsukaguchi-Honmachi, Amagasaki, 661-8661, Japan
[3]Aerospace Research Center, Japan Aerospace Exploration Agency, 6-13-1 Osawa, Mitaka, 181-0015, Japan

ABSTRACT

An 1.5 micron pulsed Coherent Doppler LIDAR system using all fiber optical components has attracted attention for remote wind sensing application because of its eye-safety, reliability and easy deployment. We report on our key technologies such as fiber based MOPA (Master Oscillator Power Amplification) transmitter, high peak power optical amplification and a real-time signal processing. Some performance results and applications are also provided.

INTRODUCTION

A Coherent Doppler LIDAR (CDL) is an attractive sensor for wind sensing because it offers a method of remote wind speed measurement in the clear atmospheric condition. The CDL involves irradiations of coherent light and detection of weak backscattered from natural aerosols in the distant atmosphere, and provides the line-of-sight component of wind speed via Doppler frequency shifts from backscattered light. This technique was first demonstrated in 1970s, and since then it has been the focus of many research projects for a variety of applications. All-fiber CDL using 1.5 micron wavelength has many advantages including eye-safety, reliability under various environmental conditions, and flexibility for components' layouts [1,2]. Mitsubishi Electric has been involved in development of a commercial prototype of an all-fiber CDL system since 2004, and has successfully confirmed its highly accurate performance based on simultaneous measurements with a conventional ultrasonic anemometer [3]. In 2005 a commercial product of an all-fiber CDL system (LR-05FC series) was manufactured [4], which successfully achieved reductions not only of its dimensions but also for manufacturing cost less than 50% of the prototype system [5].

Our all-fiber CDL has mainly used fiber-optic components which have largely improved its cost, compactness and reliability in telecommunications industry. However some specific points needed to be considered for the pulsed CDL system: generating optical seed pulses with a high on/off extinction ratio, obtaining high peak power using fiber amplification, and optical heterodyning without any fluctuation in the polarization state. Non-linear optical effects in optical fibers have to be specially considered to improve measurement performances of all-fiber CDL. Stimulated Brillouin Scattering (SBS) is one of the most dominant non-linear effects to limit measuring distance because output peak power has to be kept at SBS threshold to avoid parasitic oscillations imposing damages on fiber components.

High speed signal processing is also an important aspect for improvement of real-time processing systems because of the wide bandwidth of ~100MHz for detection of Doppler shifts at 1.5micron wavelength.

In this paper we describe the main characteristics in our standard all-fiber CDL product, some key technologies we have developed, and give an example of application to an airborne CDL for detecting Clear Air Turbulence (CAT).

STANDARD ALL-FIBER CDL SYSTEM

System configuration

Figure 1 shows a sketch of the all-fiber CDL system as well as an example of the quick look screen in the anemometer mode. The system consists of a main container (53(W) x65(H) x56(D) cm, 46kg), and an optical antenna (15(W) x15(H) x30(D) cm, 7kg) connected with optical fiber cables, and a tripod (2kg). The main container includes a fiber-based optical transmitter/ receiver (T/R) unit, a PC based signal processing unit, and its power supply in a 19 inch 10U rack case with casters which enables us to easily carry it as well as quickly set it up (~5minutes) at observation sites. The total power consumption is less than 400VA, which makes it possible to be operated by an automobile's power outlet with a commercially available DC-AC converter. By changing the scanning scheme the system has four measuring modes as follows: Line Of Sight (LOS) wind velocity with a fixed beam direction, sectored (±20deg) Plan Position Indicator (PPI) mode with a horizontal linear beam scan, sectored Range Height Indicator (RHI) mode with a vertical linear beam scan, and Velocity Azimuth Display (VAD) mode with a conical beam scan (±10deg). In this standard CDL the user interface has been considerably improved with simultaneously refreshed quick-looking screens (LOS velocity, PPI, RHI, Doppler spectra, and wind vector) as well as touch-panel user interfaces. Figure 1 (b) shows an example of the screen image (anemometer mode), which indicates the horizontal wind speed direction, and the vertical wind speed with respect to the measuring range. The refresh rate of these screens is up to a few Hz depending on the scanning rate.

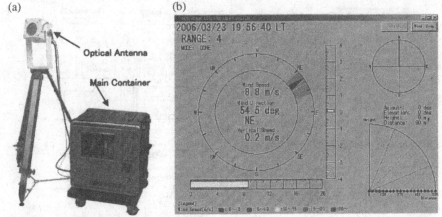

(a)

Optical Antenna

Main Container

(b)

Figure 1. A sketch of the standard all-fiber CDL system. (a) Outer view (b) An example of a quick look screen (Anemometer mode).

Optical transmitter and receiver (TRX) unit

Figure 2 shows the block diagrams of the standard all-fiber CDL system. The optical TRX unit combines a MOPA (Master Oscillator Power Amplifier) transmitter with an optical heterodyne receiver in which all the optical components are connected by polarization maintaining fiber. An 1.5 micron Distributed Feed Back Laser Diode (DFB-LD) was used as a master laser with linewidth of 0.8 MHz. Part of the optical power from the DFB-LD is used as the seed light for pulse amplification, while the remainder is used as the local power for optical heterodyning in the shot noise limited condition. The seed light is transmitted through an Acousto Optic Modulator (AOM) to introduce the frequency shift, and then modulated as burst pulses. Teeth-of a saw like waveform within a single-pulse duration were applied to the Erbium Doped Fiber Amplifier (EDFA) in order to avoid steep power inputs which may cause Stimulated Brillouin Scattering (SBS). Note that a single stage AOM was operated in a double-pass optical layout which enabled us to obtain as high on/off extinction ratio (~82dB) as a double stage AOM with half of the cost as well as power consumption. A +23dBm PM-EDFA was operated in pulse mode with optical pulse power of 10W, pulse duration of 200, 500 and 1000ns, an Pulse Repetition Frequency (PRF) of 4kHz ~16kHz without onset of SBS in pulse shape. The range resolution can be selectable from 30, 75, and 150m by changing the pulse width. These optical pulses were transferred through a fiber optic circulator to an optical antenna unit and finally transmitted to air. The backscattered light was brought back again through the same antenna, then detected as beat signals by using a balanced receiver. Finally the signals were down-converted to Intermediate Frequency (IF) signals of 54±50MHz which correspond to the Doppler velocity of ±38.5m/s.

37

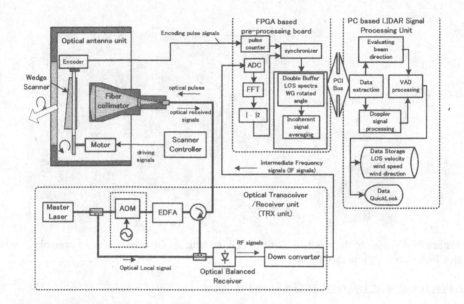

Figure 2. The block diagram of the standard all-fiber CDL system.

Optical antenna unit

The optical antenna unit consists of a fiber collimator and a compact scanner which are covered by a waterproof case. The fiber collimator has been designed as the refractive-type fiber collimator with an effective aperture diameter of 60 mm, which can set focusing from 100 m to 1 km without any truncation. It is noted that the passive-athermal design enables us to keep the Wave Front Error to less than 1/14 of the wavelength (Marechal criterion) over the temperature range of 20±30 deg., allowing operation outside the laboratory without any special care. A compact scanner enabled output beams to be conically scanned by simply rotating a silicon wedge prism and directly monitoring its rotating angle. The angular information of output beams were obtained by combining an attainable wedge-rotating angle with a constant wedge-deflecting angle. Furthermore the optional mode has been readily prepared for linear scanning by inserting an extra wedge prism reversely rotated to a primary one.

Real-time signal processing unit

Figure 3 (a) shows the schematic block diagram of a PC based signal processing unit. In this signal processing unit, a Field Programmable Gate Array (FPGA) pre-processing board has been newly developed for achieving high speed Doppler signal processing after A/D acquisition as shown in figure 3 (b). This pre-processing board contains Fast Fourier Transform (FFT), data accumulation, and coupling with encoder data as a function of beam direction from the scanner unit. In order to accommodate a bandwidth of 100MHz corresponding to a Doppler velocity of +/- 38m/sec, input IF signals are sampled at rate of 216MS/s by the A/D converter with a resolution of 8bit in the front-end of the preprocessing board. Then these sampled data are executed as 256-point FFT with appropriate zero-padding for the range resolution simultaneously for each range as Doppler spectra in a single Pulse Repetition Interval (PRI), and also incoherently accumulated up to 16000. This design enables use of range gates up to 1.19μs corresponding to a range resolution of ~150m, leading to a flexible operation for range resolutions by only changing the zero-padding length parameter.

As for the number of processing ranges, 80-ranges Doppler spectra can be attainable in real-time at PRF of 4 kHz. The number of processing ranges and PRFs, have a trade-off relationship; operation at PRF of 16 kHz can be also sustained by reducing the number of ranges to 20 in a PRI. This enables us to meet different PRF requirements for optical pulse sources in the standard CDL system as well as the middle range CDL system.

The cumulative power spectra data with each beam direction is transferred to the double buffer as 32bit x 256 x 20 data and extracted by the LIDAR data processing program via the PCI bus. In the main program spectral peak detection, Doppler velocity has been calculated after subtracting the noise floor offset for every range gate. From this processing the Line Of Sight (LOS) velocity, velocity variance and detectability are evaluated. Combining each LOS velocity with beam azimuth/elevation data, wind velocity and direction data can be estimated by using Velocity Azimuth Display (VAD) algorithms. By changing the scanning scheme the system has four measuring modes as follows: Line Of Sight (LOS) wind velocity with a fixed beam direction, sectored (±20deg) Plan Position Indicator (PPI) mode with a horizontal linear beam scan, sectored Range Height Indicator (RHI) mode with a vertical linear beam scan, and Velocity Azimuth Display (VAD) mode with a conical beam scan. The refresh rate of these screens is up to a few Hz depending on the scanning rate. It is also noted that real-time monitoring of Doppler spectra enables us to quickly comprehend whether the observation could be correctly performed on site. Furthermore measurement results are simultaneously stored in the disk drive as time series data files with a spread-sheet format, and are easily displayed back again by loading these files.

(a)

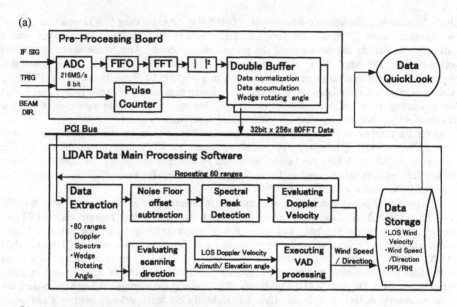

(b)

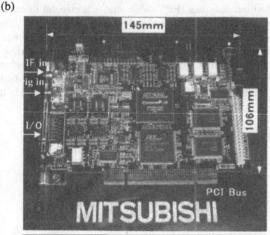

Figure 3. Schematic block diagram of a real-time signal processing unit.
(a) Schematic block diagram, (b) outer view of the FPGA based pre-processing board.

MIDDLE RANGE ALL-FIBER CDL SYSTEM

Stimulated Brillouin Scattering (SBS) is known as the dominant limiting factor of measurable distance in the all-fiber CDL. The SBS makes output optical pulses back-scatter to a fiber amplifier then propagat again as amplified pulses which may impose damages on the fiber optic components. Because of this, the transmitting pulse power has to be kept to less than the SBS threshold of 10~15W in our conventional all-fiber CDL product. In general, the SBS threshold increases with larger mode area and shorter fiber length. Therefore, we choose the short-fiber length without a Large Mode Area fiber in which the signal and pump light operate at single mode propagation in rare earth ion highly doped fibers as well as the other fiber and fiber-optic components. We believe this approach may be suitable for realization of a compact and robust optical transmitter because of the all-fiber optical circuit without any free-space coupling, leading to a long-range mobile CDL. Moreover we chose 1480nm laser diodes with single-mode fiber output as pumping sources. It was reported that Yb ions were not optically active for the wavelength of 1480nm but performed the role of a spacer for Erbium(Er) ions, inhibiting Er clustering in high concentrations of Er ions[7].

System configuration

Figure 4 shows a system configuration of the new all-fiber CDL. This system combines an optical TRX of a standard all-fiber CDL with a high peak power fiber amplifier as a power booster. In figure 4 real bold lines indicate polarization maintaining fibers and broken bold lines indicate the non-polarization maintaining single mode fibers.

A master oscillator has been replaced by a Distributed Feed Back Fiber Laser (DFB-FL) with a linewidth of 33 kHz to obtain narrower linewidth. Output pulses from a first stage EDFA with a peak power of about 10W were applied to the second AOM (AOM2) to reject the ASE (Amplified Spontaneous Emission) mainly generated in the pulse off state, and then used as seed pulses for a second stage fiber amplifier.

The second stage fiber amplifier consists of a 3.5-meter long Er3+/Yb3+ doped fiber (EYDF) based gain block with WDM (Wavelength Division Multiplexing) fiber optic couplers and single-mode Laser Diode (LD) modules for pumping. The pumping LD module has a wavelength of 1480nm at an output power of ~500mW, which was used for co-pumping, the other two were combined for counter-pumping. The EYDF has the core diameter of 9.6 micron close to that of standard single mode fiber, which allows direct fusion splicing to WDM fiber optic couplers with splicing losses less than 0.2dB. This EYDF enabled seed pulses to be amplified up to its peak power of 120W at PRF of 1 kHz without onset of SBS in their pulse shapes (SBS threshold was measured to be ~125W).

After the EYDF the output pulses were split in two orthogonal linear polarization pulses by a polarization beam splitter (PBS). The reflected pulses in the PBS have been used as monitoring signals for the state of polarization (SOP) of the EYDF which were fluctuated in the non-polarization maintaining EYDF. The SOP of the EYDF was stabilized by a fiber optic polarization controller with feed back control to minimize monitor signals. The SOP stabilized pulses were transmitted through the polarization maintaining fiber optic circulator, then collimated by a telescope with an aperture diameter of 110mm and finally transmitted to air. In order to reduce the effective fiber length of the transmitting pulse, the EYDF, PBS and the fiber optic circulator were closely mounted to the telescope in the optical antenna unit.

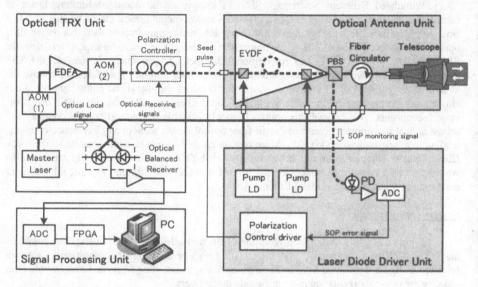

Figure 4. System configuration of the new all-fiber CDL.

Bold lines indicate the polarization maintaining fibers and broken bold lines indicate the non-polarization maintaining single mode fibers.

<u>**Experimental results**</u>

Figure 5 shows the temporal shapes of the output pulse from the EYDF gain block operated at a PRF of 4 kHz as well as that of seed pulses. The pumping powers were 500 mW for co-pump, and 1 W for counter-pump. In figure 5, the output peak energy of 58 micro joules, pulsewidth of 600 nsec and peak power of 90W were evaluated. This result indicates that the middle range CDL has 9 times more transmitting power than the standard all-fiber CDL.

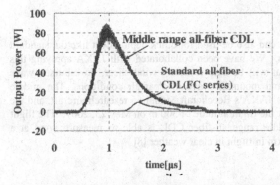

Figure 5. Optical pulse shapes of the new and conventional all-fiber CDL system.

In order to evaluate the system performance of the new CDL, the Line Of Sight (LOS) wind velocities were measured. The measuring conditions were as follows: Elevation angle of 10 deg, the incoherent integrating number of 3600, telescope focusing of infinity, and ranging resolution of 150 m. The aerosol conditions were measured to be ~9000 counts per 0.047L which were a little bit more than the average count in Kamakura (6000 counts).

Figure 6 shows the detectability depending on the distance for LOS wind velocity measured by the new CDL (plots) as well as a theoretical calculation (real curve) for the standard all-fiber CDL. This result indicates that the new CDL correctly measures LOS velocity up to 8 km because of the larger detectability than the detection limit of 4.5dB, which is 5.3 times as long as the measurable distance of the conventional all-fiber CDL. It is clearly verified that the experimental Signal to Noise Ratio (SNR) improvements of 14.7dB of the new CDL system correspond to a 1.8 times larger diameter telescope and 9 times more transmitting optical power.

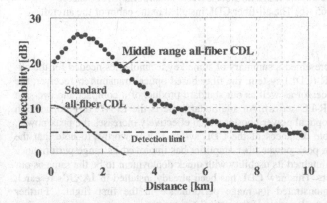

Figure 6. The comparison of the range performance between the middle-range all-fiber CDL and the standard all-fiber CDL.

Application to airborne CAT sensor

The middle range all-fiber CDL can also be used for quick deployment like the standard CDL system in an airborne installation. We have been collaborated with JAXA applying this CDL to a practical onboard sensor that will be able to detect wind shear, downbursts, wake vortex, Clear Air Turbulence (CAT) and mountain wave in clear air conditions. The middle range all-fiber CDL was installed in the JAXA's Beechcraft Model 65 research aircraft, and the first flight experiment was carried out at the flight altitude of 300 m on May 23, 2007. The flight experiment demonstrated that the middle range all-fiber CDL is able to measure wind at a maximum distance of approximately 6 km in flight in clear weather [8].

(a)   (b)

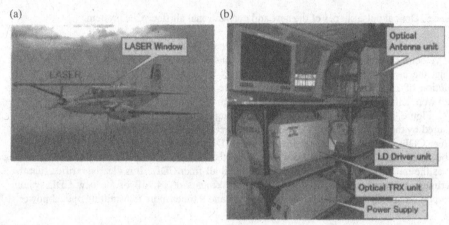

Figure7. An application of the middle range all-fiber CDL to an airborne CAT sensor.
(a) JAXA's Beechcraft Model 65 (b) The all-fiber CDL installed in the cabin of the aircraft.

SUMMARY

In this paper we have presented a summary of some key technologies needed for the all-fiber Coherent Doppler LIDAR (CDL) system: the fiber based optical transmitter/ receiver, the high speed Doppler signal processor as well as our standard product of a compact and eye-safe wind sensing CDL system (LR-FC series). An additional topic is the Er^{3+}/Yb^{3+} doped high peak power fiber amplifier as optical power booster which effectively increases the peak power, leading to improved measurable distances of the CDL system. It is worthy to note that the middle range CDL with a high peak power fiber amplifier has improved its range performance for wind sensing as well as maintained its usability with quick deployment to be the same as our standard all-fiber CDL product. This new CDL has been already installed in JAXA's research aircraft, and successfully demonstrated its range performance in the first flight. Further development for enhancing the peak power of the optical transmitter is now ongoing to improve the range performance at high altitude by Mitsubishi Electric in collaboration with JAXA.

REFERENCES

1. S. Kameyama, T. Ando, K. Asaka and Y. Hirano, *Appl. Opt.* **46,** 1953 (2007).
2. K. Asaka, T. Yanagisawa and Y. Hirano: *Proc. 10th CLRC*, 198,(1999).
3. M. Furuta, T. Matsuda, T. Ando, T. Nakayama, Jyoji Otsu and A. Komiyama: *Proc. 13th CLRC*, 251 ,(2005).
4. Y.Hirano, *CLEO2004*, CMDD4 (invited).
5. T. Ando, M. Furuta, H. Tanaka, T. Matsuda, M. Nagashima, S. Kameyama and Y. Hirano , *Proc. 23rd ILRC*, **1**, 259, (2006).
6. T. Ando, S. Kameyama, T. Sakimura, K. Asaka and Y. Hirano, *Proc.14th CLRC* (2007).
7. K. Aiso, Y. Tashiro, T. Suzuki and T. Yagi, *Furukawa Review*, **20**, 41, (2001).
8. H. Inokuchi, H. Tanaka and T. Ando, *Proc. 26th ICAS*, (2008).

REFERENCES

Mater. Res. Soc. Symp. Proc. Vol. 1076 © 2008 Materials Research Society   1076-K04-06

Linear FMCW Laser Radar for Precision Range and Vector Velocity Measurements

Diego F. Pierrottet[1], Farzin Amzajerdian[2], Larry Petway[2], Bruce Barnes[2], George Lockard[2], and Manuel Rubio[2]
[1]Coherent Applications, Inc., 101-C Research Drive, Hampton, VA, 23666
[2]NASA Langley Research Center, MS 468, Hampton, VA, 23681

ABSTRACT

An all fiber linear frequency modulated continuous wave (FMCW) coherent laser radar system is under development with a goal to aide NASA's new Space Exploration initiative for manned and robotic missions to the Moon and Mars. By employing a combination of optical heterodyne and linear frequency modulation techniques [1-3] and utilizing state-of-the-art fiber optic technologies, highly efficient, compact and reliable laser radar suitable for operation in a space environment is being developed. Linear FMCW lidar has the capability of high-resolution range measurements, and when configured into a multi-channel receiver system it has the capability of obtaining high precision horizontal and vertical velocity measurements. Precision range and vector velocity data are beneficial to navigating planetary landing pods to the pre-selected site and achieving autonomous, safe soft-landing.

The all-fiber coherent laser radar has several important advantages over more conventional pulsed laser altimeters or range finders. One of the advantages of the coherent laser radar is its ability to measure directly the platform velocity by extracting the Doppler shift generated from the motion, as opposed to time of flight range finders where terrain features such as hills, cliffs, or slopes add error to the velocity measurement. Doppler measurements are about two orders of magnitude more accurate than the velocity estimates obtained by pulsed laser altimeters [4]. In addition, most of the components of the device are efficient and reliable commercial off-the-shelf fiber optic telecommunication components. This paper discusses the design and performance of a second-generation brassboard system under development at NASA Langley Research Center as part of the Autonomous Landing and Hazard Avoidance (ALHAT) project.

INTRODUCTION

The motivation behind the development of an all-fiber high-resolution coherent lidar system comes from the need to meet requirements set by NASA's space exploration initiative. To support activities related to planetary exploration missions, this effort addresses the call for advancement of entry, descent, and landing technologies. Future exploratory missions to the Moon and Mars will become more focused towards landing at locations with high scientific value. This may include targeting sites near cliffs, valleys, craters, or other geographically interesting terrain [5,6]. Exploring technologies that will lead to an efficient and rugged method that provides feedback to the navigation and terrain hazard avoidance systems for soft landing at the targeted landing site are the main goals of this investigation [7]. This paper presents the work in progress of an all fiber lidar system capable of providing critical descent range and velocity of the planetary vehicles with a high degree of precision. A brassboard system designed to fly aboard a helicopter is currently under development for a scheduled flight test in August of 2008 over the California desert.

LASER WAVEFORM DESCRIPTION

The lidar obtains high-resolution range and velocity information from a frequency modulated-continuous wave (FMCW) laser beam whose instantaneous frequency varies linearly with time. Fig. 1 shows the transmitted (green) and received (blue) linearly chirped triangular modulation function.

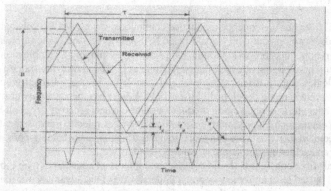

Figure 1. Frequency content of the transmitted and received linear FM-CW waveforms and the resulting time varying IF of the homodyne signal.

In a homodyne configuration, a portion of the transmitted beam serves as the local oscillator (LO) for the optical receiver. Mixing the LO field with the time delayed received field at the detector yields a time varying intermediate frequency (IF) as shown by the red trace in Fig. 1, and which is directly related to the target range by the equation

$$f_{if} = \frac{4RB}{Tc} \qquad (1)$$

where R is the range to target, B is the modulation bandwidth, T is the waveform period and c is the speed of light. For the case of a moving target, a Doppler frequency shift will be superimposed to the IF, that is a change in frequency over the waveform up-ramp ($f_{if}^{+} = f_{if} + f_{d}$) and a decrease during down-ramp ($f_{if}^{-} = f_{if} - f_{d}$). Therefore in presence of a Doppler shift, the frequency f_{if} used in Eq. (1) for determining the target range is replaced by

$$f_{if} = \frac{f_{if}^{+} + f_{if}^{-}}{2} \qquad (2)$$

The Doppler frequency shift for the case when the shift is less than f_{if} is simply

$$f_d = \frac{f_{if}^+ - f_{if}^-}{2} \qquad (3)$$

Target radial velocity component is obtained from the equation

$$v = \frac{f_d \lambda}{2\cos\theta} \qquad (4)$$

where λ is the transmitter laser wavelength, and θ is the angle between the target velocity vector and the lidar line of sight.

Horizontal and vertical vector velocity components are obtained by dividing the transmit laser power into three beams and angularly separating each laser beam from each other by 120 degrees in azimuth, and approximately 45 degrees in nadir. Having prior knowledge of the orientation of the beams relative to the platform direction of motion allows us to measure horizontal as well as vertical velocities without ambiguities. In addition, since there are still two solutions for each set of horizontal and vertical frequency components the sensor can distinguish between positive or negative Doppler frequency shifts by keeping track of the waveform's ramp up and ramp down temporal locations. For example, consider the case of a positive Doppler shift during forward motion; the IF between the received signal and the LO during the up-ramp will be less than the IF during the down ramp as is displayed in Fig. 1. If the motion is in the opposite direction, the up-ramp IF will be greater than the down-ramp IF. By comparing the IF of the up-ramp to the IF of the down-ramp, the Doppler ambiguity is removed for each of the three transmitted beams.

SYSTEM DESCRIPTION

The performance quality of the device comes primarily from the line width of the seed laser and the linearity of the modulated waveform. Laser line-width limits the IF measurement resolution, and waveform linearity limits signal to noise ratio (SNR) [8, 9]. The system to be installed and field-tested aboard a helicopter uses a frequency stabilized, fiber laser with a spectral line width that is less than 10 kHz full width, at half maximum. A great deal of effort was devoted to achieving a linear modulation and transform limited signal spectra operation. Fig. 2 shows the measured laser frequency vs. time illustrating nearly perfect waveform linearity.

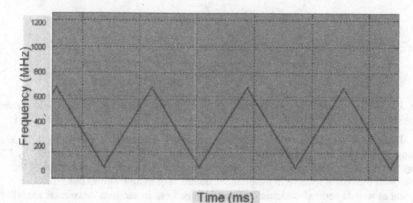

Figure 2. Spectrogram shows the frequency modulated waveform, which was obtained by optically mixing the lidar waveform with a stable continuous wave reference laser.

Fig. 3 is a block diagram that illustrates the configuration of the all fiber FMCW lidar sensor. The FMCW waveform generated at the seed laser is amplified by a high power fiber amplifier. Line width measurements at the output of the amplifier show negligible line broadening by the amplifier. The output of the fiber amplifier is split into three components in order to distribute the power to three telescopes at the sensor's optical head. The optical head consists of the fiber to free space coupling telescopes, which can be mounted separately from the rest of the system anywhere on the vehicle platform that has a clear view of the target. The signals from the ground are collected by these telescopes and sent to three pairs of dual balanced detectors via optical fiber cables. The outputs of the detectors are processed by a Pentium Dual Core processor based receiver, capable of either storing the temporal data for post processing, or providing real-time range and velocity measurements that are currently displayed on a graphical user interface (GUI). Real-time range and velocity measurement is a critical capability necessary for precision navigation. All system electronics and fiber optic components are housed in a standard instrumentation rack.

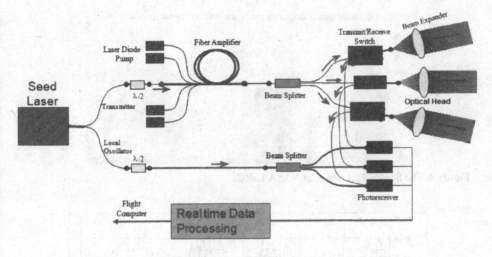

Figure 3. All fiber, coherent Doppler lidar system schematic.

SAMPLE DATA PRODUCTS

Performance characterization of the lidar is ongoing at the ALHAT Sensor Test Range (STR) at NASA Langley Research Center. The sensor test range includes a target board with diffuse surfaces of known reflection coefficients and different size panels at a 250-meter range [10]. A photograph of the STR is shown in Fig. 4 for reference. Range characterization experiments of the FMCW lidar were made on the right section of the STR consisting of 30-cm wide by 2-meter long panels that extend away from the backboard. The distance between the panels and the backboard range from 25-cm to 5-cm from top to bottom, in steps of 5-cm increments, and one panel is set at 2.5-cm at the bottom for a total of six panels. To test the range measurement performance of the FMCW lidar, the fiber-coupled telescopes are mounted on a rotating stepping motor that scans the laser beam along a vertical path over the target board. Fig. 5 shows the ranges calculated from the measured signal IF per Eqn. (1). Superimposed on the measured data are the actual range values.

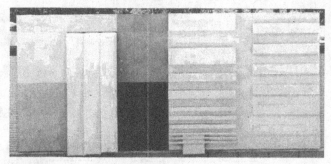

Figure 4. The Sensor Test Range at NASA LaRC

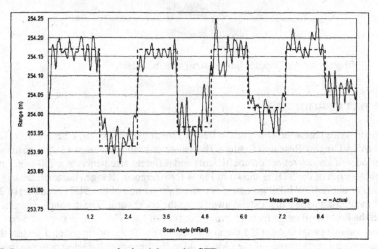

Figure 5. Range measurements obtained from the STR.

The panels shown in Fig. 5 correspond to the 25-cm panel at the left, followed by the 20-, 15- and down to the 10-cm panel on the right of the chart. This preliminary data collected by the real-time system shows that the measured range values agree well with the true values to less than 5-cm. Ongoing improvements are expected to improve the range measurement accuracy to about 1-cm.

To generate velocity signals, a rudimentary wheel was constructed and mounted on an electric motor. The perimeter surface of the wheel consists of a rough sandy texture uniformly painted with a color of known diffuse reflectance at the lidar wavelength. The wheel has a 91-cm diameter, and a perimeter surface width of 15-cm. Fig. 6 shows the measured velocity with respect to time at a position on the wheel that is approximately 45-degrees up from normal incidence (normal incidence is the line of sight angle pointed to rotation axis of the wheel). As the data was collected, velocity variations were detected and displayed by the real-time data acquisition and processing system. The periodic variations of the target velocity were caused by wheel wobble, and the sudden change in the velocity is due to a wind gust pushing on the wheel, as pointed out in Fig. 6.

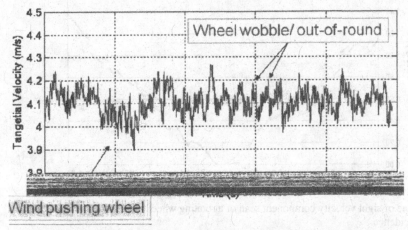

Figure 6. Velocity measurements of wheel obtained from 250-meters standoff range.

Velocity measurements were also made by scanning the lidar beam from above the normal incidence angle on the wheel to below the normal incidence angle, in order to obtain varying line-of-sight velocities and changing magnitude from positive to negative. Fig. 7 shows a partial scan of the wheel velocity distribution above and below normal incidence. Discontinuities appearing on the velocity measurements are once again due to wind and wheel wobble.

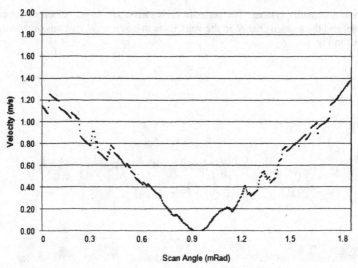

Figure 7. Line of sight velocity component scan of a rotating wheel. Zero velocity corresponds to normal incidence.

SUMMARY

A precision range and velocity lidar sensor is under development at NASA LaRC. Key features of this lidar are all-fiber design, narrow laser linewidth operation, excellent waveform modulation linearity, and eye-safe operation. The preliminary characterization tests indicate precisions of the order of 1-cm in range and 1-cm/sec in velocity measurements. This lidar is currently being integrated into a relatively compact, rugged package for a series of tests from a helicopter platform. These helicopter tests are planned in support of NASA's Autonomous Landing and Hazard Avoidance Technology (ALHAT) project.

REFERENCES

1. Skolnik, M.I., "Introduction to Radar Systems" 2nd ed., (McGraw-Hill Book Company, New York, 1980).
2. Saunders, W.K., "Post War Developments in Continuous-Wave and Frequency Modulated Radar," *IRE Trans.*, vol ANE-8, pp. 7- 19, March 1961.
3. A.L. Kachelmyer, "Range-Doppler imaging: wave-forms and receiver design," *Proc. SPIE*, **999**, 138-161 (1988).
4. Jelalian, A.V., "Laser Radar Systems," (Artech House, Massachusetts, 1992).
5. Space Studies Board, National Research Council, New Frontiers in the Solar System – An Integrated Exploration Strategy, (National Academy Press, Washington, D.C., 2003).

6. Golombek, M.P., Cook, R.A., et al, "Overview of the Mars Pathfinder Mission and Assessment of Landing Site Predictions," Science Magazine, **278**, 1743–1748, December 5, 1997.
7. Wong, E.C., et al., "Autonomous Guidance and Control Design for Hazard Avoidance and Safe Landing on Mars", AIAA Atmospheric Flight Mechanics Conference and Exhibit 5-8, 4619, Monterey, California, August 2002.
8. Pierrottet, D.F., et. al., "Development of an All-Fiber Coherent Laser Radar for Precision Range and Velocity Measurements", Materials Research Society, San Francisco, California 2005.
9. Karlsson, C.J, et. al., "Linearization of the frequency sweep of a frequency-modulated continuous-wave semiconductor laser radar and the resulting ranging performance", Appl. *Opt.*, **38** (15), May 1999.
10. Pierrottet D.F., et. al., "Characterization of 3-D imaging lidar for hazard avoidance and autonomous landing on the Moon", *Proc. SPIE*, Orlando, Florida, April 2007

Mater. Res. Soc. Symp. Proc. Vol. 1076 © 2008 Materials Research Society　　　　1076-K04-07

Simplified Homodyne Detection for Linear FM Lidar

Peter Adany[1], Chris Allen[2], and Rongqing Hui[3]

[1]ITTC, University of Kansas, 247N Nichols Hall, 2335 Irving Hill Rd., Lawrence, KS, 66045
[2]CReSIS, University of Kansas, 321 Nichols Hall, 2335 Irving Hill Rd., Lawrence, KS, 66045
[3]ITTC, University of Kansas, 222 Nichols Hall, 2335 Irving Hill Rd., Lawrence, KS, 66045

ABSTRACT

A fiber based lidar system is developed which simplifies the processing of linear FM pulses by using a modulated local oscillator power in the coherent detector. Experiments were conducted on lidar systems with direct, heterodyne and simplified homodyne detection to compare receiver sensitivity. A field experiment using the homodyne system verified the sensitivity estimation on a building target at 370-m range.

INTRODUCTION

The need to better understand and predict climate change requires more efficient and compact high resolution instruments to map ice topography from satellites and small aircraft. Compared to classic radars, fiber based lidar systems offer narrower radiating angle, higher target contrast and higher resolution. Combined with pulse compression techniques, lidar can provide high performance using off-the-shelf components from the communications industry.

In traditional pulsed lidar systems, a simple transmit and receive scheme is used that achieves range resolution proportional to the pulse duration as

$$\Delta R = c\tau/2 \qquad (1)$$

where τ is the pulse duration and c is the speed of light. To achieve fine range resolution, short pulses are used and thus high pulse peak powers are needed to preserve the signal-to-noise ratio in the receiver. The maximum transmit power is limited by the hardware, causing limited range resolution in these systems. Furthermore, in q-switched lasers the emitter contains damagingly high optical fields that limit the system's lifespan to less than 5 years.

In previous research at the University of Kansas a lidar system was demonstrated that uses linear FM pulse compression to achieve fine range resolution with low peak power [1]. This system applies RF modulation to the optical pulse, in the form of a linear frequency sweep (chirp). A unique feature of this method is that recovering the range information simply requires multiplying the received RF chirp signal with a local reference chirp. This system achieved heterodyne detection using a wavelength-tuned laser as a local oscillator at an intermediate frequency of several GHz.

In the heterodyne system, high fidelity analog RF circuitry is required after the optical receiver to process the linear FM chirp, resulting in a hybrid system with a relatively complex architecture. However, a shortcut is possible because both linear FM and homodyne detection use time domain multiplication of signals. In the simplified homodyne design the coherent receiver is supplied with a modulated local oscillator so that the photodetector directly produces a dechirped RF signal. The block diagram of this system is shown in Figure 1 below.

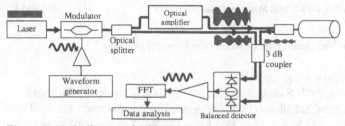

Figure 1. Block diagram of the homodyne system.

This arrangement simplifies the optical and RF receiver circuits compared to heterodyne detection as it bypasses down conversion from the IF band and subsequent baseband dechirping. A second advantage of this is the substantial reduction of electrical bandwidth required in the photodiode and receiver circuitry.

EXPERIMENT

In order to validate this principle and compare detection sensitivity, a lidar test bed was assembled with direct, heterodyne and simplified homodyne detection. First the three systems were tested using an ideal transmission channel simulated by a 22 km fiber spool and optical attenuator.

Following the ideal channel tests, a 5"-aperture diameter Newtonian telescope was incorporated into the homodyne system. A paper target was placed indoors at 50-m distance, and a free space power meter was used to measure the incident power. Lastly, an echo was recorded from a building target at approximately 370-m range as shown in Figure 2, allowing us to verify the sensitivity and range accuracy predictions.

Figure 2. Building target at 370-m. An aerial image from Google Maps is shown at left (with line of sight marked by a blue line). A photo from the telescope view point is shown at right.

RESULTS

The measured sensitivity was compared against theory and computer simulation. The homodyne system's sensitivity approaches the quantum limit and surpasses our comparable heterodyne and direct detection systems, as shown in Figure 3.

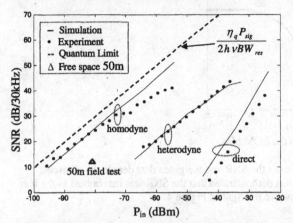

Figure 3. SNR comparisons of three detection methods.

In the above figure, the circles indicate experimental data and its respective simulation results for comparison.

In the homodyne system, it was found beneficial to bias the optical modulator in a technique similar to duobinary modulation [2]. This mode doubles the modulation frequency and reduces the transmitted power between pulses. The comparison between standard and doubled biasing modes is illustrated in Figure 4.

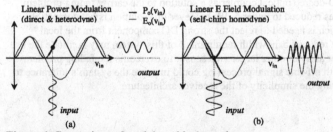

Figure 4. Comparison of modulator biasing points.

We verified this effect by altering the input voltage DC bias as shown in Figure 5.

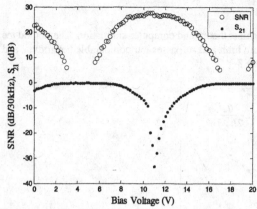

Figure 5. SNR and DC power versus the EOM bias.

In the above figure the blue circles denote the SNR and the green dots denote the quiescent output power of the modulator. This test demonstrated that the SNR was maximized as the bias level reached the minimum point shown in the right of Figure 4.

DISCUSSION

While the heterodyne and homodyne systems should provide similar quantum-limited sensitivity, the heterodyne system was significantly worse in the experiments and the computer simulation. We attribute this to the IF down conversion process which degrades the signal to noise ratio [3].

One challenge with the homodyne design is its low tolerance to optical phase fluctuation due to environmental effects such as vibration, atmospheric distortion, and Doppler shift. This is a typical problem in homodyne receivers, as a 90 degree phase mismatch of the 1310 nm wavelength laser results in destructive interference of the received signal. We tested a phase diversity receiver using a 90-degree optical coupler as a solution to the carrier fading problem. The magnitude of fading was reduced to under 5-dB; however this solution is not compatible with balanced detection which is needed to reject the strong DC component from the local oscillator. Nevertheless, the reduced bandwidth requirement of this system may facilitate the use of a large area photodetector array or other larger area sensor to overcome the fading problems. Such a design combined with off-line signal processing could improve the system's tolerance to carrier fading while preserving the simplicity of the receiver architecture.

CONCLUSIONS

A simplified lidar system was developed that integrates linear FM pulse compression with a homodyne optical receiver. This system provides quantum noise limited sensitivity in a compact architecture, and its theory has been validated by computer simulation, lab bench experiment and free space trials. One major challenge is signal fading due to optical phase fluctuation. This issue may be resolved by designing a free space optical receiver using a focal plane array or other large area detector arranged in a balanced configuration.

ACKNOWLEDGMENTS

This research is supported by the University of Kansas Center for Remote Sensing of Ice Sheets (CReSIS) and Information and Telecommunications Technology Center (ITTC).

REFERENCES

1. C. Allen, Y. Cobanoglu, S. K. Chong and S. Gogineni, "Performance of a 1319 nm Laser Radar Using RF Pulse Compression," Proceedings of the 2001 International Geoscience and Remote Sensing Symposium (IGARSS '01), Sydney, session SS52, paper Th03-04, pp. 997-999, July 2001.
2. K. Yonenaga and S. Kuwano, "Dispersion-Tolerant Optical Transmission System Using Duobinary Transmitter and Binary Receiver," IEEE J. Lightwave Technology, Vol. 15, pp. 1530-1537, 1997.
3. M. Schwartz, W. R. Bennett and S. Stein, "Communication Systems and Techniques", (IEEE Press 1996), ISBN 0-7803-4715-3 (Chapter 3).

Fiber Optic and Semiconductor Lasers

Mater. Res. Soc. Symp. Proc. Vol. 1076 © 2008 Materials Research Society 1076-K01-02

Microstructured Soft Glass Fibers for Advanced Fiber Lasers

Axel Schulzgen[1], Li Li[1], Xiushan Zhu[1], Shigeru Suzuki[1], Valery L. Temyanko[1], Jacques Albert[2], and Nasser Peyghambarian[1]

[1]College of Optical Sciences, University of Arizona, 1630 E. University Blvd., Tucson, AZ, 85750
[2]Department of Electronics, Carleton University, 1125 Colonel By Drive, Ottawa, K1S 5B6, Canada

ABSTRACT

Combining novel highly-doped phosphate glasses and advanced fiber drawing techniques, we fabricated and tested single-frequency fiber lasers that generate powers of more than 2 W. We demonstrate enhanced performance employing active photonic crystal fiber compared to more conventional devices that are based on large core step-index fiber.

We also present results on phase-locking and coherently combining the output of up to 37 fiber cores into a single, near-Gaussian laser beam. To achieve exclusive oscillation of the fundamental in-phase supermode, several all-fiber laser cavities have been designed, numerically analyzed, fabricated, and tested. We will report on a 10 cm long monolithic all-fiber laser that emits more than 12 W of optical power and is based on combining the output of 19 active cores. All the cores are integrated within the same cladding and arranged in a two-dimensional isometric array. Our truly all-fiber approach that omits any free-space optical elements lead to a multi-emitter laser device that is free of optical alignment and robust against external perturbations.

INTRODUCTION

Almost all commercial fiber lasers use rare earth doped silica optical fiber as the active material. Combined with fiber Bragg gratings (FBGs) that provide the appropriate optical feedback at the emission wavelength monolithic, all-fiber lasers can be fabricated with low-loss fusion splices between the different silica fiber components. On the other hand, doped phosphate glasses appear to be the best active material for many bulk lasers. These glasses allow for extremely large doping levels with negligible clustering effects. Er/Yb co-doped phosphate glasses exhibit the additional advantage of having very high phonon energy and, consequently, a high energy-transfer efficiency from absorbing Yb ions to emitting Er ions. During recent years compact fiber lasers based on Er/Yb co-doped phosphate glass fiber have been demonstrated [1-4]. They can be forced into single-frequency operation [1, 4] for applications including LIDAR and coherent optical sensing but also made quite powerful with W-level output [2, 3] to fabricate building blocks for laser systems with scalable optical power. Several advances in phosphate fiber fabrication including microstructured fibers and photosensitive fibers have enabled recent progress in phosphate glass fiber laser devices.

Here we will discuss two novel approaches to increase the active emission volume of compact phosphate glass based fiber lasers utilizing advanced drawing techniques to fabricate microstructured optical fiber (MOF). All phosphate MOFs have been drawn at the College of Optical Sciences, The University of Arizona. First, we show that very large active mode areas can be achieved applying photonic crystal type active fibers with regular patterns of air holes in the cladding that run along the fiber length. These large mode areas lead to improved absorption in short length of active fiber and, therefore, result in increased output powers of compact and single-longitudinal mode fiber lasers. Secondly, we show that the emission volume can also be increased in discrete steps by incorporating several doped cores into the active fiber section. We will demonstrate that an all-fiber technique can be applied to coherently combine the emission of several cores into a single beam with a near-Gaussian beam profile.

RESULTS AND DISCUSSION

Fiber lasers with large mode area microstructured fiber

The output power of a single mode fiber laser is mainly limited by the small core size due to the single mode criteria that restrict the core diameter. E.g., in the step-index fiber a 13 μm diameter core of our previous phosphate fiber laser [2] is already considered large compared with those of most commercial silica single mode fibers (≤ 8 μm). Since the doping concentration can only be increased until the detrimental ion clustering effect occurs, mode size increase is the only available option to increase the number of active ions that participate in the lasing process while maintaining single mode guidance. Within the recently developed concept of MOF single mode guidance is demonstrated with mode areas much larger than that achievable by step index fibers [5-7]. Therefore, this concept originally applied in silica fiber provides a promising solution to expand the active single mode core area for Er^{3+}-Yb^{3+}-codoped phosphate fiber lasers.

We have fabricated phosphate glass based MOF and two different fiber laser cavities shown schematically in figure 1. These have been studied. Both of them utilize a MOF segment with a large, highly doped phosphate glass core to provide high pump absorption and large optical gain. A microscope image of the active fiber cross-section is shown in figure 1 (c). The MOF has an outer diameter of 125 μm, a pitch (Λ, center-to-center spacing of neighboring air holes) of 9 μm, and a doped central area of ~430 μm^2 corresponding to a core diameter of 23 μm. The MOF has four rings of air holes surrounding the core, and the air hole diameter (d_{AH}) varies when drawn under different temperature. The index of the core glass is depressed by an amount of $\Delta n = n_{core}$-$n_{cladding} = -1.7 \times 10^{-3}$ to achieve SM guidance [8, 9]. When d_{AH}/Λ reaches 0.4, the MOF guides only one spatial mode and the measured M^2 value of this mode is ~1.2. The deviation from the ideal value of 1.0 of a diffraction-limited beam is because the fundamental mode pattern of the MOF differs from the Gaussian mode. The MOF core is doped with 1.5 wt.% Er_2O_3 and 8 wt.% Yb_2O_3. This MOF exhibits very high absorption of highly multimode 975 nm pump light that is coupled into the MOF cladding. Less than 10% of the pump light is transmitted through the first 50 mm of the active MOF.

In the cavity shown in figure 1 (a) the pump light is coupled through a dielectric mirror with high transmission at the pump wavelength (975 nm) and high reflection at the signal wavelength (~1535 nm) into the 50 mm long active fiber [10]. An additional output coupler is used with an optimized reflectivity of 54% for the signal wavelength and 100% reflectivity at 975 nm to recycle the unabsorbed pump and further improve the laser efficiency.

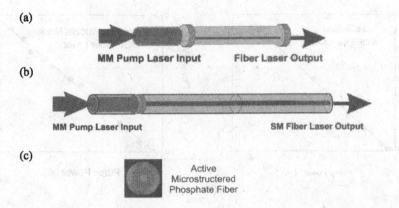

(a)

MM Pump Laser Input   Fiber Laser Output

(b)

MM Pump Laser Input   SM Fiber Laser Output

(c)

Active
Microstructered
Phosphate Fiber

Figure 1. Schematic of the two different fiber laser configurations (a) and (b) that utilize active microstructured phosphate glass fiber. In configuration (a) two broadband dielectric coatings are used to form the laser cavity, while in configuration (b) a broadband mirror on the pump side is combined with a Narrow band fiber Bragg grating to achieve single-longitudinal mode laser emission. In (c) a microscope image of the active microstructured fiber with 125 microns outer diameter and heavily rare earth doped core is shown.

The performance of this ultra compact fiber laser is shown in figure 2 (a). The output power reaches more than 5 W and the power per length has been greatly improved to 1.34 W/cm compared to previous step index fiber lasers, e.g., it has been more than double compared to a previously reported 0.56 W/cm from a phosphate glass step index fiber laser [2]. This clearly demonstrates the advantage of the MOF design over the conventional SIF in improving the device compactness as well as compressing the active fiber length sufficiently short for single frequency laser operation.

To achieve single-longitudinal mode operation a narrow band highly diffractive element has to be introduced into the laser cavity. Here we introduced a FBG with narrow band reflection as an output coupler into the fiber laser cavity [11] as illustrated in figure 1 (b). All other cavity parts including the input coupler and the active MOF are the same as in the previously described fiber laser. At the fiber laser output side the active fiber is spliced to a single mode silica fiber (Nufern PS-GDF-20/400) that has a large area photosensitive core. The nominal core diameter is 20 μm with a numerical aperture of 0.06. The original outer diameter of this fiber is 400 μm. To achieve a low loss fusion splice between the active phosphate PCF and the photosensitive silica fiber the latter has been etched to an outer diameter of 125 μm by hydrofluoric acid. The FBG is written into the core of this silica fiber using 244 nm light and a 25 mm long phase mask. The reflection spectrum of the FBG shows a peak reflectivity of 17% at 1.534 μm and a 3 dB bandwidth of about 0.03 nm. Per roundtrip the complete fiber laser has a propagation loss of 3.8 dB due to the fusion splices and coupling losses between different fibers and the output loss is 7.7 dB. The roundtrip loss can easily be compensated by the large gain in the PCF with the highly doped, large area core.

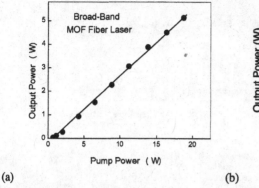

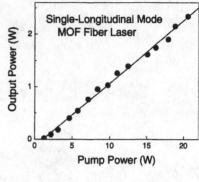

(a)   (b)

Figure 2. Performances of two very short MOF fiber lasers. In (a) results for a configuration illustrated in figure 1 (a) with 50 mm of active MOF are shown, while in figure (b) a single-longitudinal mode laser with FBG (see figure 1 (b)) is tested. This fiber laser consists of only 38 mm of active MOF.

The output versus pump power characteristics of this single-longitudinal mode MOF laser is shown in figure 2 (b). A slope efficiency of about 12% is observed up to pump powers of 20 W. In contrast, previously reported short length, single-frequency lasers showed strong saturation effects at similar pump levels [4]. The slightly lower efficiency compared to the broad-band laser in figure 2 (a) is mainly due to additional cavity losses due to the intracavity splicing joint between phosphate and silica fiber. This effect can be reduced in the future using recently demonstrated FGBs in phosphate glass fiber [12]. A maximum output power over 2.3 W demonstrates the advantages of utilizing a large core MOF over conventional step-index fibers and indicates the potential of cladding pumped single-frequency fiber lasers.

Fiber lasers with multiple active cores

Active fibers with multiple doped cores is a promising approach to a discrete increase in fiber laser mode area and have the potential to provide a power-scaling solution to compact high-power fiber laser devices [13]. By splitting the gain medium into discrete positions inside the cladding, instead of concentrating all active ions in an oversized core, thermal issues are of less concern for multicore fiber (MCF) and the optical power extracted per unit fiber length can be increased [14]. However, there still exists a challenge to effectively obtain high-brightness output beams from MCF lasers to make this design advantageous compared with the conventional single-core fiber, namely, the proper selection of the oscillating supermode. Usually, in a MCF with a 2-D isometric array, if each core is single-mode and neighboring cores are optically coupled, the MCF has a total of non-degenerated supermodes equal to the number of cores. Among all supermodes, only the in-phase mode (all cores emit in phase) has the preferential far-field intensity distribution (Gaussian-like) [15]. Therefore, the resonator needs to be explicitly designed to selectively establish the in-phase mode oscillation. Several coherent beam combining techniques are commonly applied to phase lock multiple emitters, including Talbot-cavity, structured mirror, collimating lens with high reflector and self-Fourier transform resonator. All these conventional approaches involve free-space optics, i.e., air spacing and bulk optics, as part of the laser cavity. The inclusion of free-space optics not only increases the device size and alignment complexity substantially, but also decreases the laser efficiency owing to the additional cavity losses. More importantly, it also causes serious instability problems during high power laser operation because of its susceptibility to environmental and thermal disturbances accompanied with increased power level. To take full advantage of the fiber laser device, i.e., the compact size as well as the robustness against external disturbances, the phase locking operation should ideally take place inside a confined waveguide. Based on our advance fiber fabrication technique several MCF have been fabricated with up to 37 cores. As an example, a 19-core fiber (MC19) is shown in figure 3 (a). The MC19 cores are again made from heavily Er/Yb co-doped phosphate glass, of 1.5 wt.% Er_2O_3 and 8.0 wt.% Yb_2O_3 [16]. The MC19 has an outer diameter of 200 μm and a pump-confining inner cladding diameter of 110 μm. Each individual core has an effective diameter of 7.6 μm and a NA of 0.12 at 1.55 μm, and it is single mode with a full angular spread of 0.26 rad. We have calculated the supermodes of this MC19 using a finite element method with the near- and far-field intensity distributions of the in-phase supermode. The far-field distribution shown in figure 3 (b) has a full angular spread of 0.04 rad, corresponding to a 47-μm-diameter effective waist for the Gaussian envelope of the in-phase mode.

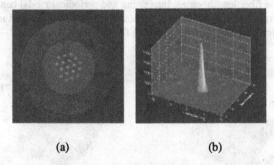

(a)   (b)

Figure 3. Microscopic image of the active 19-core fiber (a) with calculated nearly Gaussian far-field (b) intensity distributions of the fundamental in-phase supermode.

In the first experiment, a short piece of MC19 (10 cm in length) is used as the gain fiber. A fiber laser was built by a bare piece of MC19 without any modal control elements. By butt-coupling one end of the MC19 against a multimode pump-delivery fiber that had a broadband high-reflector (at ~1.5 μm) coated on its facet, the 975 nm pump light was launched into the MC19 cladding. The MC19 started to lase at a pump threshold of ~4 W and the output far-field distribution was recorded on a screen 7.5 cm away from the output end. The far-field pattern is highly irregular as shown in figure 4 (a), with a full angular spread of ~0.19 rad. Please notice that this angle is narrower than that of a single emitter, meaning that the emitters are coupled and the output beam is a mixture of several of the 38 supermodes.

Next we show that by combining Talbot and diffraction effects, the preferable in-phase supermode can be selected in an all-fiber device configuration [16]. Passive fibers with low transmission loss are spliced to the output ends of a MCF. The passive fiber had uniform index over its cross-section, i.e., it was coreless, and a diameter of 200 µm that was matched to MC19 to reduce splice loss. A relatively long piece of passive fiber, ~1.7 mm, was spliced to the output end and another shorter piece of ~ 0.5 mm length was spliced to the pump end. This fabricated device was pumped and tested in the same way as the first one, with the resulting far-field pattern shown in figure 4 (b). A clean and well-confined on-axis spot is observed with a horizontal spread angle of 0.04 rad, which is identical to the predicted value of the in-phase supermode shown in figure 3 (b). The beam spot is a little elongated in the vertical direction and we believe the imperfections of the manufactured MC19, e.g., noncircular core shapes and facet cleaving defects as seen in figure 3 (a) are responsible. In all, figure 3(e) is predominantly distinguished from 3(a), which clearly demonstrates the effectiveness of our all-fiber supermode selection approach.

Simply speaking, segments of passive fiber with well selected length can be used to confine, propagate, reflect and recouple the waves in order to select the in-phase supermode and discriminate higher order modes. The resulting fiber laser is a monolithic all-fiber device that can be operated at output power levels above 10 W.

It should be pointed out that the lengths of the spliced passive fibers at both ends of MCF are not chosen arbitrarily. It is not necessarily true that longer mode-selecting fiber leads to better results. As the passive fiber length varies, interesting and periodical modal behaviors have been observed, resulting from combined effects of Talbot imaging, diffraction and multimode interference for sufficiently large and long passive fibers [17].

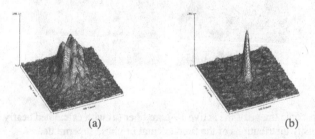

(a)   (b)

Figure 4. Measured far-field intensity distributions of the 19-core fiber laser emissions: (a) bare 19-core fiber without supermode selection and (b) with an all-fiber supermode selection technique [16] applied that enables exclusive in-phase supermode oscillation.

This all-fiber approach can be applied to MCFs with different outer diameters and arrays with various quantities of cores. A recent study on the 37-core fiber [18] highlights its tremendous application potentials for high-power high-brightness laser devices. Here, we would like to note that the potential of MCF lasers goes beyond extracting more power per unit length of fiber. The freedom of designing the multiple core arrays into various patterns allows exploiting beam steering and shaping techniques, in analogy to phased array RADAR technology.

As an example we studied an MCF with 12 cores that are arranged in a rectangular array. Rectangular array MCF lasers are expected to exhibit particular near-field distribution, far-field diffraction pattern, and splitting of propagation constants, which are different from those of other MCF arrays according to the array modal analysis using coupled-mode theory. Our 12-core fiber shown in figure 5(a) not only has 12 cores but also has air holes that run along the fiber. The 12 cores are arranged in a 3×4 rectangular array. The diameter of individual doped-cores is 8.5 μm and that of the air holes is about 2 μm. The pitch of the microstructure is 8 μm and that results in the periods of the core array to be 14 μm and 16 μm in x and y directions, respectively. The refractive index of the core is 1.5698 and that of the cladding is 1.5690. This microstructure results in 12 individual cores with a numerical aperture of 0.16 and a modal field diameter of 11 μm. Therefore, the effective modal area of the whole structure is 1140 μm^2. The individual cores are co-doped with 1 wt% Er_2O_3 and 2 wt% Yb_2O_3, respectively. The MCF has an outer diameter of 125 μm which enables directly end-pumping by a multimode fiber coupled diode laser. Figure 5 (b) and (c) show the near-field and far-field intensity distributions of the array during in-phase supermode operation. They very closely resemble the theoretically calculated supermode patterns [19].

The interesting point is the ability to design a particular supermode property. Here, the in-phase supermode is split into two non-degenerated modes corresponding to two polarization states. The frequency spacing between the two longitudinal modes is 120 MHz and the corresponding propagation constant birefringence $\Delta\gamma$ is about 4×10^{-6} 1/μm, i.e., we could design and clearly observe the effective group index birefringence of about 10^{-6}. This polarization mode discrimination can be attributed to differences in the propagation constant β of individual cores and in the coupling coefficient for the two polarization fields. These differences are induced by the asymmetry of the rectangular-shaped array and the microstructure of the MCF.

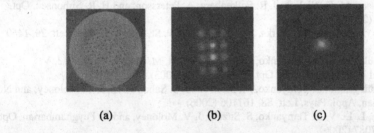

(a)   (b)   (c)

Figure 5. Microscopic images of the output facet of a 12-core microstructured Er/Yb co-doped phosphate fiber (a) and measured near-field (b) and far-field (c) intensity distributions during in-phase supermode operation [19]. By design the in-phase supermode of this fiber is birefringent.

CONCLUSIONS

We have demonstrated that the ability to fabricate active microstructured fibers increases the degree of freedom in designing fiber optic devices and becomes an efficient tool to build compact and efficient fiber lasers. Using this fabrication technique we improved the performance of single-longitudinal fiber lasers and designed features of the emitted laser mode such as shape and polarization properties. Microstructured fiber enables tailoring fiber optic components and devices to specific applications.

ACKNOWLEDGMENTS

The authors would like to thank E. Temyanko and Y. Merzlyak for technical support. This work was supported by the National Sciences Foundation through grant No. 0725479, the Natural Sciences and Engineering Research Council of Canada through grant SROPJ 334867-2005, and the state of Arizona TRIF Photonics Initiative.

REFERENCES

1. Ch. Spiegelberg, J. Geng, Y. Hu, Y. Kaneda, S. Jiang, and N. Peyghambarian, J. Lightwave Tech. **22**, 57 (2004).
3. T. Qiu, L. Li, A. Schülzgen, V. L. Temyanko, T. Luo, S. Jiang, A. Mafi, J. V. Moloney, and N. Peyghambarian, IEEE Photon. Technol. Lett. **16**, 2592 (2004).
2. L. Li, M. M. Morrell, T. Qiu, V. L. Temyanko, A. Schülzgen, A. Mafi, D. Kouznetsov, J. V. Moloney, T. Luo, S. Jiang, and N. Peyghambarian, Appl. Phys. Lett. **85**, 2721 (2004).
4. T. Qiu, A. Schülzgen, L. Li, A. Polynkin, V. L. Temyanko, J. V. Moloney, and N. Peyghambarian, Opt. Lett. **30**, 2748 (2005).
5. J. C. Knight, T. A. Birks, R. F. Cregan, P. St. J. Russell, and J. -P. de Sandro, Electron. Lett. **34**, 1347 (1998).
6. P. St. J. Russell, Science **299**, 358-362 (2003).
7. N. A. Mortensen, M. D. Nielsen, J. R. Folkenberg, A. Petersson, and H. R. Simonsen, Opt. Lett. **28**, 393 (2003).
8. B. J. Mangan, J. Arriaga, T. A. Birks, J. C. Knight, and P. St. J. Russell, Opt. Lett. **26**, 1469 (2001).
9. L. Li, A. Schülzgen, V. L. Temyanko, H. Li, S. Sabet, M. M. Morrell, A. Mafi, J. V. Moloney, and N. Peyghambarian, Opt. Lett. **30**, 3275 (2005).
10. L. Li, A. Schülzgen, V. L. Temyanko, M. M. Morrell, S. Sabet, H. Li, J. V. Moloney, and N. Peyghambarian, Appl. Phys. Lett. **88**, 161106 (2006).
11. A. Schülzgen, L. Li, V. L. Temyanko, S. Suzuki, J. V. Moloney, and N. Peyghambarian, Opt. Express 14, 7087 (2006).
12. J. Albert, A. Schülzgen, V. L. Temyanko, S. Honkanen, and N. Peyghambarian, "Strong Bragg gratings in phosphate glass single mode fiber," Appl. Phys. Lett. **89**, 101127 (2006).
13. P. Glas, M. Naumann, A. Schirrmacher, and T. Pertsch, in *Technical Digest of Conference on Lasers and Electro-Optics* (Institute of Electrical and Electronics Engineers, New York, 1998), pp. 113.
14. Y. Huo and P. K. Cheo, IEEE Photon. Technol. Lett. **16**, 759 (2004).
15. Y. Huo, P. Cheo, and G. King, Opt. Express **12**, 6230 (2004).
16. Li, A. Schülzgen, S. Chen, V. L. Temyanko, J. V. Moloney, and N. Peyghambarian, Opt. Lett. **31**, 2577 (2006).

17. Wrage, P. Glas, M. Leitner, T. Sandrock, N. N. Elkin, A. P. Napartovich, and D. V. Vysotsky, Proc. SPIE **3930**, 212 (2000).
18. Li, A. Schülzgen, H. Li, V. L. Temyanko, J. V. Moloney, and N. Peyghambarian, JOSA B **24**, 1721 (2007).
19. X. Zhu, A. Schülzgen, L. Li, H. Li, V. L. Temyanko, J. V. Moloney, and N. Peyghambarian, Optics Express **15**, 10340 (2007).

Mater. Res. Soc. Symp. Proc. Vol. 1076 © 2008 Materials Research Society   1076-K01-04

Efficient High Power ns Pulsed Fiber Laser for Lidar and Laser Communications

Jian Liu

Fiber Laser Research, PolarOnyx, Inc., 470 Lakeside Drive, Suite F, Sunnyvale, CA, 94085

Abstract

Research results on high energy/power ns pulsed fiber lasers are discussed in this paper. Modulation schemes for a seed laser in getting various optical waveforms, high power operation of fiber amplifiers, nonlinearity mitigation in high power fiber lasers, and trade-offs have been addressed. It shows experimentally that ns pulses can be achieved by direct modulation of a semiconductor seed laser and an energy level of sub-mJ were demonstrated. 23 % wall plug efficiency was achieved for a compact laser module.

Introduction

Figure 1 High power fiber lasers is an enabling technology for NASA deep space communications and Lidar applications

For future NASA's deep space missions and Lidar applications, high energy pulsed fiber lasers have been considered to be an enabling technology to build high power transmitters [1-10]. The next generation pulsed fiber laser requires a weight of smaller than 1kg, high pulse repetition rate (> 10 MHz for communications and > 10 kHz for Lidars), ns pulse width, > 10 W average output power and > 1 kW peak power at 1.06 micron and 1.55 micron, and overall efficiency higher than 20%. In addition to those, high extinction ratio (>30 dB), high OSNR, and good pulse shape are highly desired for enhancing system performance to battle with attenuation, pulse distortion, and pulse delay towards longer distance of transmission [4-7]. These requirements eliminate classic Q-switch fiber lasers. Different amplification schemes can be used as alternatives to produce high energy and high repetition rate pulses via a gated amplification, and multiple isolated gain stages.

In this paper, various types of single frequency nanosecond fiber laser will be presented. Up to 20 W operation and sub-mJ fiber lasers were achieved with over 23% wall plug efficiency.

Experimental Results on Pulse Shaping Fiber Laser

Figure 2 plots the experimental setup for the demonstration of pulse shaping technology in the fiber laser. The checking points were labeled A through E in Figure 2. Both spectrum and pulse shape were taken at those checking points to evaluate the pulse evolution within the fiber laser. Pulse widths from 500 ns to 1 ns and repetition rates from 10 kHz to 60 MHz were obtained from our experiment by tuning the driving electronics of the seed laser.

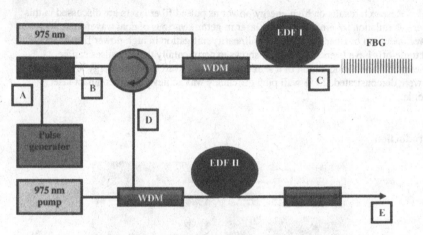

Figure 2 Experimental setup for a pulse shaping fiber laser

Figure 3 gives examples of Gaussian shape pulse generation and propagation in the fiber laser. The Gaussian shape pulse was generated by changing the driving current format and bias of the direct modulation DFB laser (the repetition rate can be tuned from 10 kHz to 60 MHz and the pulse width from 1 ns to 500 ns). Strong pulse shaping effects are shown in Figure 3 for Gaussian shape amplification at a repetition rate of 30 MHz. The pulses maintained their shape through all stages of amplifiers when using the FBG filter. However, when taking out the FBG, the pulse becomes distorted and degrades into a square shape pulse at the output (checking point E). 4 ns pulses show a more distorted shape than 8 ns pulses. The direct modulation laser driven under threshold current can obtain a pulse with a high extinction ratio over 30 dB. Drawings of amplitude vs. time in Figure 3 (a) and (b) were taken from the output signal at checking point E by an oscilloscope and shows such effect. In Figure 3 (a), a drawing of the pulse from checking point B is also given for a comparison. There is no difference for the pulse shape between B and E when a FBG filter is used.

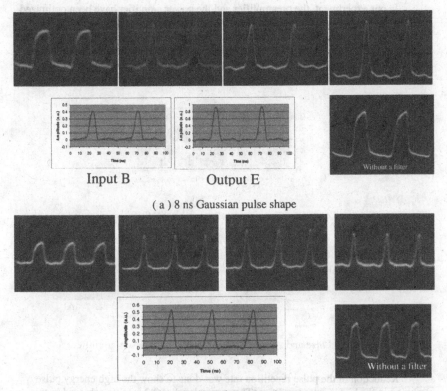

Input B   Output E

(a) 8 ns Gaussian pulse shape

(b) 4 ns Gaussian pulse shape

Figure 3 Gaussian shape pulse propagation in the fiber laser (30 MHz)

Figure 4 shows a typical spectrum evolution for pulse propagation in the high power fiber laser. It is shown that by using a FBG filter, the output spectrum can be cleaned up to get over 45 dB OSNR by rejecting out of band ASE and side lobe spectrum of direct modulation laser. Even though it is 25 dB below the signal, the side lobe (about 1.3 nm separation from the main signal wavelength) causes XPM that contributes to the pulse distortion in the high power amplifier stage of the fiber laser, as shown in Figure 3.

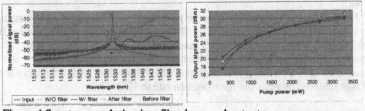

Figure 4 Spectrum evolution in a fiber laser and output power vs pump power

In our experiment, the preamplifier and the power amplifier have been optimized and characterized as well for the fiber laser. The pump conversion efficiency can be as high as 38%. This laser has a flat spectrum for different wavelengths at 1550 nm band. Furthermore, the pulse shaping fiber laser can be controlled to achieve excellent pulse shape stability and power stability. Figure 5 shows the power stability of the pulse shaping fiber laser. 3 % power fluctuation is obtained for over forty consecutive pulses.

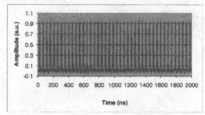

Figure 5 A train of pulses showing excellent stability of power and pulse shape

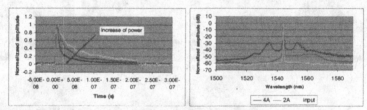

Figure 6 Pulse and spectra evolution in the high power amplifier

Reduction of the pulse repetition rate was done to show the high energy pulse evolution in the high power amplifier. The energy level of 0.2 mJ was achieved in this experiment at 1550 nm. One important phenomena was observed in this experiment is that the pulse was compressed significantly when it evolves through the amplification stage. A 200 ns pulse width was compressed down to 10 ns at the highest energy level. This is partially due to the gain dynamics and partially due to the soliton formation in the anomalous dispersion fiber. By manipulating the pulse shape launched into the amplifier, it can be resolved. Further investigation is on-going in our lab. Figure 6 shows the pulse evolution and spectra at different pump currents. Further energy scaling can be done by use of the spectral shaping technology with multimode fiber or photonic crystal fiber so the large mode field diameter can accommodate the high peak power and reduce further on the SBS.

For a fiber laser operating at a center wavelength of 1064 nm, the performance is similar to that of 1550 nm. A packaged 10 W fiber laser operating at 1064 nm has been delivered to NASA JPL and a 20 W fiber laser to MDA. Figures 7 & 8 show its performance of wall plug efficiency and beam profile. The wall plug efficiency was measured to be 25% for a full power operation and M2 is below 1.05. An energy level of 0.5 mJ was achieved by reducing the repetition rate to 10's kHz.

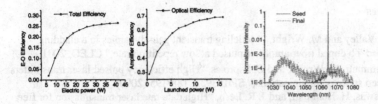

Figure 7 Performance of a packaged 10 W fiber laser: wall plug efficiency (left), pump power conversion efficiency (middle) and spectra of input and output (right)

Figure 8 10 W and 20 W fiber laser systems and Beam profile

Summary

In summary, we have demonstrated a pulse shaping and spectral shaping fiber laser to achieve down to 4 ns pulse. The average power is over 1 W at 1550 nm (energy 0.2 mJ) and over 20 W at 1064 nm (energy 0.5 mJ). Over 30 dB extinction ratio and 45 dB OSNR have been achieved as well. Excellent pulse shape stability and power stability is also demonstrated. Wall plug efficiency of 23% was achieved in hardware delivery. Further shaping of the waveformat will be feasible to scale up the energy to over mJ by reducing the pulse narrowing effect for long ns pulse and distortion for short ns pulse during amplification.

Acknowledgement
This paper is supported in part by NASA and MDA through SBIR contracts. The author thanks Dr. Farzin Amzajerdian (LARC, NASA), Dr. Malcolm Wright (JPL, NASA) and Mr. Fred Robinson (SMDC) for their continuous help and encouragement.

References

1. G. C. Valley and M. Wright, "Modeling transient gain dynamics in a cladding pumped Yb doped fiber amplifier pulsed at low repetition rate," CLEO, 2001.
2. H. Hemmati, M. Wright, and C. Esproles, "High efficiency pulsed laser transmitters for deep space communications," SPIE 3932, 188-195 (2000).
3. A. Biswas, H. Hemmati, and J. R. Lesh, "High data rate laser transmission for free space laser communications," SPIE 3615, 269-277 (1999).
4. J. M. Elmirghani and R. A. Cryan, "Performance analysis of self synchronized optically preamplified PPM, " J. Lightwave Technol. 13 (5), 923-932 (1995).
5. R. W. Ziolkowski and Justin B. Judkins, "Propagation characteristics of ultrawide bandwidth pulsed Gaussian beams," J. Opt. Soc. Am. A 9(11), 2021-2030 (1992).
6. D. Kelly, C. Y. Young, and L. C. Andrews, "Temporal broadening of ultrashort space-time Gaussian pulses withy applications in laser satellite communication," SPIE 3266, 231-240 (1998).
7. J. B. Hartlay, "NASA's future active remote sensing missions for earth science," SPIE 4153, 5-12 (2001).
8. S. W. Henderson, et al., "Eye safe coherent laser radar for range and micro Doppler measurement," Proc. IRIS Active Systems 1997, Vol. 1, Tucson, AZ (1997).
9. http://www.gsfc.nasa.gov/
10. James Abshire, Laser sounder technique for remotely measuring atmospheric CO_2 concentrations. James.B.Abshire@nasa.gov.

Mater. Res. Soc. Symp. Proc. Vol. 1076 © 2008 Materials Research Society

Electrically Pumped Photonic Crystal Distributed Feedback Quantum Cascade Lasers

Manijeh Razeghi, Yanbo Bai, Steven Slivken, and Wei Zhang
Department of Electrical Engineering and Computer Science, Northwestern University, Center for Quantum Devices, 2220 Campus Drive, Evanston, IL, 60208

ABSTRACT

In parallel with the effort to improve the efficiency of Quantum cascade lasers (QCL) for high power continuous wave (CW) operations, the peak power in pulsed mode operation can be easily scaled up with larger emitting volumes, i.e., processing QCLs into broad area lasers. However, as the emitter width increases, multi-mode operation happens due to poorer lateral mode distinguishability. By putting a two dimensional photonic crystal distributed feedback (PCDFB) layer evanescently coupled to the main optical mode, both longitudinal and lateral beam coherence can be greatly enhanced, which makes single mode operation possible for broad area devices. For PCDFB laser performance, the linewidth enhancement factor (LEF) plays an important role in controlling the optical coherence. Being intersubband devices, QCLs have an intrinsically small LEF, thus serving as better candidates over interband lasers for PCDFB applications. We demonstrate electrically pumped, room temperature, single mode operation of photonic crystal distributed feedback quantum cascade lasers emitting at $\lambda \sim 4.75$ µm. Ridge waveguides of 50 µm and 100 µm width were fabricated with both PCDFB and Fabry-Perot feedback mechanisms. The Fabry-Perot device has a broad emitting spectrum and a broad far-field character. The PCDFB devices have primarily a single spectral mode and a diffraction limited far field characteristic with a full angular width at half-maximum of 4.8 degrees and 2.4 degrees for the 50 µm and 100 µm ridge widths, respectively.

INTRODUCTION

High power, single mode, mid-infrared laser sources are highly desirable for a number of applications, such as countermeasures, free space communication and remote chemical detection. Besides the constant effort devoted to improving the light generation mechanism within the emitting core materials, which fundamentally governs the capability of high power operation, novel and superior laser cavities need to be developed to ensure robust single mode operation. Although the wall plug efficiency of the current state-of-art quantum cascade lasers (QCL) reaches close to 10% [1], the single mode maximum output power is limited by the lack of a proper laser cavity.

Broad area QCLs are natural candidates to produce multi-watt peak power, however, as the gain width increases, multi-mode operation happens due to poorer lateral mode distinguishability. In addition, nonlinear phenomena like spatial hole burning and filamentation becomes more pronounced. As such, the beam coherence cannot be maintained and divergence far from diffraction limited angles occurs, which prevents the conventional broad area Fabry-Perot device for single mode operation.

The angled-grating DFB (α-DFB) [2] uses a tilted grating with respect to the laser facet. Due to enhanced lateral spatial coupling, filamentation can be strongly suppressed. Although α-DFB has been demonstrated capable of producing nearly diffraction limited spatial beam quality, the spectral single mode operation is difficult to achieve due to the inefficient longitudinal feedback, which is otherwise proven to be very effective in the conventional 1D-DFB [3-5].

An approach that combines both the spatial beam quality of the α-DFB and the spectral beam quality of the 1-D DFB is the photonic crystal distributed feedback (PCDFB) laser [6], which generalizes the 1-D grating geometry to a 2-D pattern, allowing single mode operation both in far field and spectrum. For PCDFB laser performances, the linewidth enhancement factor (LEF) plays an important role in controlling the optical coherence [7]. Being intersubband devices, QCLs have an intrinsically small LEF [8], which makes them better candidates over interband lasers for PCDFB applications. Optically pumped PCDFB was demonstrated on type-II interband cascade lasers (ICL) at λ ~ 4.4 μm [9]. Recently, electrically pumped PCDFB was demonstrated on 980 nm InGaAs/AlGaAs interband lasers [10]. Here, we report electrically pumped PCDFB quantum cascade lasers emitting at λ ~ 4.75 μm.

EXPERIMENT

PCDFB design

We started with the same QCL wafer as that in Ref. [3], which had a 30- stage-QCL core followed by a 10 nm InP etch stop layer and a 100 nm GaInAs grating layer. The envisioned PCDFB structure after a 4μm InP cladding regrowth is sketched in figure 1(a). A targeting wavelength of λ = 4.75 μm is chosen based on the emission spectra of the Fabry-Perot lasers in Ref. [3]. Using a 1-D effective index method, the transverse fundamental mode is calculated, resulting a refractive index contrast for the 100 nm grating layer of Δn = 0.00797. The calculated effective index for the fundamental mode is n_{eff} = 3.196, following an 1-D effective index method.

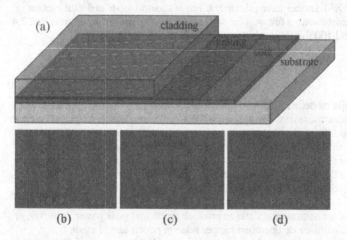

Figure 1. (a) Sketch of the PCDFB structure and SEM pictures for (b) top view of the etched SiO₂ surface, (c) oblique view of the etched GaInAs holes before the removal of SiO₂, and (d) cross section of the regrowth.

The PCDFB device in this work incorporated a first order photonic crystal lattice with circular elements and rectangular symmetry. The ratio of the lattice constants in two directions is set to be an integer number of 3. Fig. 2 sketches the primitive cell of the pattern in both real and reciprocal spaces. The lattice constants in the two directions are given by

$$\Lambda_1 = \lambda / \left(2 n_{eff} \sin \phi\right) \tag{1}$$

and

$$\Lambda_2 = \lambda / \left(2 n_{eff} \cos \phi\right), \tag{2}$$

where ϕ is the tilt angle of the 2D lattice, which reads

$$\phi = \tan^{-1}\left(\frac{1}{3}\right) \approx 18°. \tag{3}$$

The calculated periods of the first order PCDFB are $\Lambda_1 = 2.349$ μm and $\Lambda_2 = 0.783$ μm in the two primary axes, respectively. Tolerance of effective index calculation error is introduced by exposing three PCDFB gratings with $\Lambda_2 = 0.77$ μm, 0.78 μm and 0.79 μm, while keeping $\Lambda_1 = 3$ Λ_2. The later testing result shows that the first grating, instead of the second, produces room temperature operation at 4.75 μm, with the threshold from the other two longer gratings beyond the maximum current density that the devices can sustain. This indicates some underestimation of the effective index resulting from the simple 1-D effective index method. A more accurate estimation for the effective index would be $n_{eff} = 3.251$ and can be used for future processing of the same wafer.

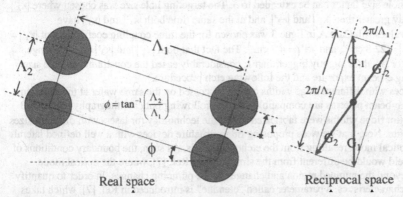

Figure 2. Sketch of the primitive cell for the 2D PCDFB pattern in both real and reciprocal space, showing the relation between the tilt angle and the lattice constants. The chosen reciprocal vectors are also shown.

Given the chosen lattice constants and fixed grating depth (100 nm), the coupling coefficients only change with the size of the holes. We define a hole size factor β, which is the ratio of the hole diameter to the smaller lattice constant Λ_2. The coupling coefficients are directly proportional to the refractive index contrast, weighted by an area integral on the 2D primitive cell, which calls for some reciprocal vectors. The expression for the original coupling coefficients is given by

$$\kappa_i = \frac{2\pi\Delta n}{\lambda} \frac{1}{\Lambda_1\Lambda_2} \iint_s ds \exp(-i\mathbf{G}_{\kappa i} \cdot \mathbf{r}), \tag{4}$$

where i reads 1, 2, and 3, with

$$\mathbf{G}_{\kappa 1} = \mathbf{G}_{-1} - \mathbf{G}_1, \tag{5}$$

$$\mathbf{G}_{\kappa 2} = \mathbf{G}_2 - \mathbf{G}_1, \tag{6}$$

and

$$\mathbf{G}_{\kappa 3} = \mathbf{G}_{-2} - \mathbf{G}_1. \tag{7}$$

In order to be consistent with Ref. [7], the original coupling coefficients are scaled with the tilt angle by

$$\kappa_i' = \kappa_i \cos\phi. \tag{8}$$

Figure 3 shows the calculated coupling coefficients κ_1', κ_2', and κ_3' as a function of the hole size factor. Note that for a photonic crystal lattice with circular elements, κ_1' and κ_3' do not vanish at $\beta = 1$, which is in distinct contrast to the case with square elements. As a result, the range of the hole size factor can be extended to 3. The targeting hole size was chosen where $|\kappa_2'|$ is significantly greater than $|\kappa_1'|$ and $|\kappa_3'|$, and at the same time both $|\kappa_1'|$ and $|\kappa_3'|$ have appreciable values [6]. Case A in figue 3 was chosen for the three coupling coefficients of $|\kappa_1'| = 3.4$ cm^{-1}, $|\kappa_2'| = 22.8$ cm^{-1}, and $|\kappa_3'| = 4.7$ cm^{-1}. The fact that both $|\kappa_1'|$ and $|\kappa_3'|$ do not change appreciably for β values slightly bigger than 1, considerably eased the constraints on hole size control during e-beam exposure and the following etch procedure.

Devices with different ridge widths can be processed on the same wafer at the same time provided the e-beam patterns are compatible with the following optical lithography pattern. 50 μm and 100 μm ridge widths were fabricated to test the technology for lasers with different sizes of gain medium. Note that we were proposing ridge structure devices with a well defined lateral size of the optical mode resulting from the etched channels. As such, the boundary conditions of the optical field would be different from the simulation in Ref. [7], where the optical mode penetrates beyond the pumped region and changes as the pumping changes. In order to quantify the far field characteristics, a parameter called "etendue" is introduced in Ref. [7], which takes into account of the changeable near field patterns. As a special case, a diffraction limited far field pattern corresponds to an etendue value of unity.

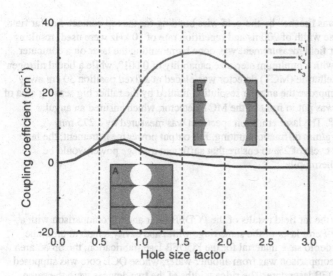

Figure 3. Calculated coupling coefficients for a rectangular photonic crystal lattice with circular elements, where $\Lambda_1 = 3\Lambda_2$, $\lambda = 4.75$ µm, and $\Delta n = 0.00797$. The hole size factor is the ratio of the hole diameter to the smaller lattice constant Λ_2.

Device processing

The PCDFB processing started with an 80 nm SiO_2 deposition on the GaInAs grating layer, followed by e-beam exposure of the designed pattern on a 100 nm thick e-beam resist. The 2D grating pattern was transferred to the SiO_2 layer by reactive ion etching (RIE). Figure 1(b) shows the top view of the pattern on a SiO_2 surface. The 100 nm GaInAs grating layer was etched by electron cyclotron resonance (ECR) RIE at a temperature of 150°C in a Cl_2/Ar plasma. Figure 1(c) shows the high magnification oblique view of the GaInAs etching results. The remaining SiO_2 was then removed by buffered HF solution, followed by a short acid treatment before loading into the metal-organic chemical vapor deposition (MOCVD) chamber for InP upper cladding regrowth, which includes a 2 µm thick n- InP layer with graded doping (1-3 x 10^{17} cm^{-3}) and a 2 µm thick n+ InP layer (4×10^{18} cm^{-3}). Shown in Figure 1(d) is the regrowth cross section, where the grating on top of the core can be clearly seen.

After regrowth, the wafer was processed into double channel ridge waveguide lasers of 50 µm and 100 µm ridge widths, using conventional photolithography and nonselective wet chemical etching. After insulation deposition, window opening, top contact deposition, lapping/polishing, and bottom contact deposition, following the same procedure described in Ref. [3], the wafer was then cleaved into laser bars with 3 mm cavity length (in a direction normal to the cleaved mirrors) and indium soldered to copper heat sinks with the epilayer side up. The device packaging was finished by the gold wire bonding for the top contact. For far field and spectrum testing, a pulse width of 200 ns and repetition rate of 10 kHz were used, resulting in a duty cycle of 0.2%. Far field measurement was done by mounting the laser on a computer controlled rotational stage with a minimum step size capability of 0.001°, while a liquid nitrogen cooled mercury-cadmium-telluride (MCT) detector was placed at a fixed position 30 cm away from the rotation axis. To improve the angular resolution limited by the rather big sensing area of the detector, a 200 μm slit was put in front of the MCT detector, which ensured an angular resolution of less than 0.05°. The laser emission spectrum was measured by a 275 mm monochromator using a 75 g/mm diffraction grating. For output power measurement, the laser was driven at a higher duty cycle of 2% to ensure that sufficient average power would be received by the calibrated thermopile.

Device testing

Shown in Fig. 4 are the far field results of the PCDFB laser and the comparison with a previously processed Fabry-Perot laser of the same ridge width and cavity length at the same driving current of 5A. In order to save material for the PCDFB laser fabrication, the broad area Fabry-Perot laser used for comparison was from another wafer, whose QCL core was supposed to be the replica of the PCDFB laser core. The ridge widths of the two devices were the same, being 50 μm. It is clearly shown that the PCDFB device has a significantly narrower far field, with a full width at half maximum (FWHM) of 4.8°, while the conventional Fabry-Perot device shows a much broader and irregular far field pattern.

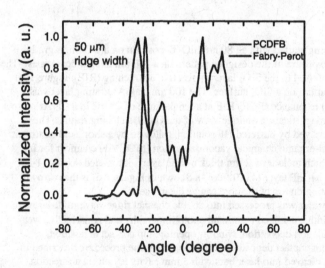

Figure 4. Experimentally measured far field pattern of a PCDFB device and a Fabry-Perot device with the same ridge width of 50 μm. The FWHM of the PCDFB device is 4.8 degrees.

Since our device incorporated a ridge waveguide structure, the lateral size of the optical field is well defined, therefore we are able to determine whether the PCDFB far field is diffraction limited or not. Both PCDFB devices with 50 μm and 100 μm ridge widths can be simulated using the same method. Taking the 100 μm ridge width device as an example, we performed the standard Fraunhofer transform

$$F(\theta) = \cos\theta \int_{x=-a/2}^{x=a/2} \exp[if(x)]\exp(-ikx\sin\theta)dx \qquad (9)$$

to a single slit with the same width

$$a = d/\cos\phi, \qquad (10)$$

as the PCDFB laser facet, where d is the ridge width (100 μm), and θ is the diffraction angle. Angle

$$\phi = \tan^{-1}(1/3) \approx 18° \qquad (11)$$

denotes the tilt of the cavity with respect to the facet normal, which traces back to the ratio of Λ_1 and Λ_2 [7]. To account for arbitrary emission angle with respect to the facet normal, the slit is assumed obliquely illuminated by a plane wave, represented by a linear phase function of

$$f(x) = kx\sin\delta \qquad (12)$$

from one side of the slit to the other side, where δ denotes the oblique angle and

$$k = 2\pi/\lambda. \qquad (13)$$

Setting the oblique angle to be the same angle as the measured far field maxima position, the simulated normalized diffraction intensity $|F(\theta)|^2 / |F(\theta = \delta)|^2$ is plotted as the blue curve in figure 5, where the only adjustable parameter is the oblique angle δ. The close resemblance of the simulation and experimental far field measurement for both the main peak and even the side lobes clearly indicates that the PCDFB far field is indeed diffraction limited. The position and the FWHM of the far field pattern are -18.4° and 2.4°, respectively. Simulation for the 50 μm ridge width PCDFB device also shows diffraction limited far field.

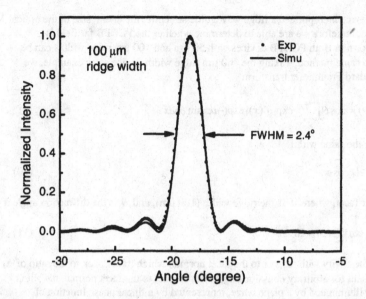

Figure 5. Experimentally measured and simulated far field pattern of a PCDFB device with a ridge width of 100 μm. The driving current is 10 A.

We notice that in our result the output beam is distinctively off the facet normal, which is in contrast to that reported in Ref. [10], where the far field maxima is in the direction normal to the facet. The interpretation given in Ref. [10] cannot explain our results due to the lack of phase information along the facet. Our results can be understood by incorporating a non-negligible phase gradient along the facet. Better understanding of the off-normal far field is being carried on.

Figure 6 shows the spectrum results of the PCDFB laser and the comparison with that of a similarly-processed Fabry-Perot laser with the same ridge width. Note that the Fabry-Perot laser results are from a similar but different QCL core, so the absolute position of the spectral features is not relevant. Figure 6 is intended to show the distinct contrast of general multimode behavior in the wide-ridge Fabry-Perot lasers and the single mode behavior of the PCDFB laser. The peak position and the FWHM of the emission spectrum read 4.755 μm and 4 nm, respectively.

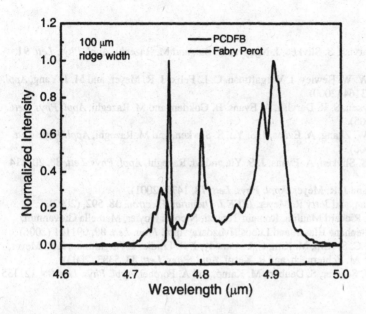

Figure 6. Comparison of the spectral characteristics between a PCDFB laser and a Fabry-Perot laser with the same ridge width.

CONCLUSIONS

Electrically pumped PCDFB quantum cascade lasers were realized with room temperature operation. The ridge widths of these devices were 50 μm 100 μm. The far field and spectrum were compared with its Fabry-Perot counterpart. Both far field and spectrum of the PCDFB laser showed single mode behavior. The far field was diffraction limited. The emitting wavelength was 4.755 μm. Being a proof of concept experiment, the output power of the PCDFB device was not optimized and the slope efficiency was about one fifth of its Fabry-Perot counterpart. Once optimized the superior beam quality of the PCDFB quantum cascade laser over its Fabry-Perot counterpart opens the possibility of making single-mode, high-power broad area mid-infrared semiconductor lasers.

ACKNOWLEDGMENTS

The authors would like to acknowledge the support, interest, and encouragement of J. Meyer from Naval Research Laboratory, H. Temkin, M. Rosker, and R. Leheny from Defense Advanced Research Projects Agency, and experts from Army Research Office and Office of Naval Research.

REFERENCES

1. Evans, S. R. Darvish, S. Slivken, J. Nguyen, Y. Bai, and M. Razeghi, *Appl. Phys. Lett.* **91**, 071101 (2007).
2. R. E. Bartolo, W. W. Bewley, I. Vurgaftman, C. L. Felix, J. R. Meyer, and M. J. Yang, *Appl. Phys. Lett.* **76**, 3164 (2000).
3. J. S. Yu, S. Slivken, S. R. Darvish, A. Evans, B. Gokden, and M. Razeghi, *Appl. Phys. Lett.* **87**, 041104 (2005).
4. S. R. Darvish, W. Zhang, A. Evans, J. S. Yu, S. Slivken, and M. Razeghi, *Appl. Phys. Lett.* **89**, 251119 (2006).
5. S. R. Darvish, S. Slivken, A. Evans, J. S. Yu, and M. Razeghi, *Appl. Phys. Lett.* **88**, 201114 (2006).
6. I. Vurgaftman and J. R. Meyer, *Appl. Phys. Lett.* **78**, 1475 (2001).
7. Igor. Vurgaftman, and Jerry R. Meyer, *IEEE J. Quantum Electron.* **38**, 592, (2002).
8. Thierry Aellen, Richard Maulini, Romain Terazzi, Nicolas Hoyler, Marcella Giovannini, Jérôme Faist, Stéphane Blaser, and Lubos Hvozdara, *Appl. Phys. Lett.* **89**, 091121 (2006).
9. W. W. Bewley, C. S. Kim, M. Kim, C. L. Canedy, J. R. Lindle, I. Vurgaftman, J. R. Meyer, R. E. Muller, P. M. Echternach, and R. Kaspi, *Appl. Phys. Lett.* **83**, 5383 (2003).
10. H. Hofmann, H. Scherer, S. Deubert, M. Kamp, and A. Forchel, *Appl. Phys. Lett.* **90**, 121135 (2007).

Mater. Res. Soc. Symp. Proc. Vol. 1076 © 2008 Materials Research Society   1076-K07-03

Development of Low-Cost Multi-Watt Yellow Lasers Using InGaAs/GaAs Vertical External-Cavity Surface-Emitting Lasers

Mahmoud Fallahi[1], Li Fan[1], Chris Hessenius[1], Jorg Hader[2], Hongbo Li[2], Jerome Moloney[2], Wolfgang Stolz[3], Stephan Koch[3], and James Murray[4]

[1]College of Optical Sciences, University of Arizona, Tucson, AZ, 85721
[2]ACMS, University of Arizona, Tucson, AZ, 85721
[3]Philipps Universitat, Marburg, Germany
[4]Arate Associates, Longmont, CO, 80501

ABSTRACT

We demonstrate a highly strained InGaAs/GaAs VECSEL operating at 1173 nm with more than 8.5 W output power and tunable over 40 nm. High-efficiency yellow-orange emission is then achieved by intra-cavity frequency doubling. Over 5 W of CW output power in the 585-589 nm spectral regions is achieved. This compact low-cost high-power yellow-orange laser provides an innovative alternative for sodium guidestar lasers or other medical / communication applications.

INTRODUCTION

Laser sources covering 570-590 nm bands are of great interest for sodium guidestar lasers, quantum computing, dermatology and ophthalmology applications [1-3]. Despite these major benefits, the development of these yellow-orange lasers has been very limited, mainly because it is hard to find active materials with direct transition in this band. Nonlinear frequency conversions have been frequently used to generate emission in the yellow-orange range. Several methods including frequency doubling of Yb solid-state lasers [4], frequencies doubling of Raman-shifted Yb (Nd) lasers [5], sum-frequency generation in solid state lasers [6] and frequency doubling of Bi-doped fiber lasers [7] have been investigated. Unfortunately the majority of these approaches suffer from limited emission range, low output power or high cost. Semiconductor lasers are very attractive for their high-gain, low-cost and large volume production capability and are widely used in the near IR range for optical communication. However due to their limited direct band-gap energy, a range of visible emission wavelengths are hard to be directly fabricated.

Optically pumped vertical-external-cavity surface-emitting laser (VECSEL) using multi-quantum well semiconductors are very attractive for low-cost high-power high-brightness sources [8, 9]. In addition, by having access to the intracavity, several attractive features such as wavelength tuning, frequency doubling for visible generation and Q-switching can be achieved.

DESIGN AND FABRICATION

Strained InGaAs/GaAs multi-quantum well lasers are widely used for the generation of 900-1000 nm lasers. Here we report on the development and demonstration of a highly strained

InGaAs/GaAs VECSEL which can cover a significantly longer wavelength range of 1147 nm - 1197 nm. Very robust multi-Watt high-brightness performance at room temperature is demonstrated. Using intracavity frequency doubling we demonstrate high-power coherent emission in a wide yellow-orange band (575 ~595 nm). This low-cost compact wavelength agile laser is an attractive candidate to replace existing yellow sources.

The design of the VECSEL structure is accomplished using rigorous many-body microscopic quantum design tools and 3D optical/thermal modeling of the device [9]. This microscopic quantum design approach, which utilizes a closed-loop semiconductor laser design tool that is free of adjustable fit parameters, rigorously computes the low-intensity photoluminescence (PL) spectra, semiconductor gain (and refractive index) spectra, spontaneous and Auger recombination losses for the specific QW structure. Computed low-intensity PL spectra are used for wafer diagnostic and quality control. Coupled with the quantum design approach, the 3D optical/thermal modeling of the device enables closed-loop design and optimization of the VECSEL semiconductor chip prior to wafer growth.

One major challenge in the development of InGaAs/GaAs VECSEL operating in the 1150 nm ~1200 nm range is the growth of highly strained InGaAs/GaAs multi-quantum well layers. The designed InGaAs/GaAs VECSEL structure was grown using a low temperature metal organic vapor phase epitaxial (MOVPE) process. The MOVPE growth uses alternative liquid MO-V-sources (tertiarybutylarsine-TBA, tertiarybutylphosphine-TBP) that decompose at lower temperatures than the conventional hydride precursors. This allows for a general reduction of the growth temperature, which promotes higher values of strain and, thus, reproducibly higher indium-concentrations in the active QW, essential for long wavelength operation. In addition, strain-compensating GaAsP barriers with precise chemical composition are grown. These factors enabled growth of InGaAs epitaxial structure that achieves lasing at the target wavelength of 1170 nm. The VECSEL structure consists of 10 layers of InGaAs compressive strained quantum wells. Each quantum well is 7-nm thick and surrounded by GaAsP strain compensation layers and AlGaAs barriers, in which the 808-nm pump emission is absorbed. The thickness and compositions of the layers are optimized such that each quantum well is positioned at an antinode of the cavity standing wave to provide resonant periodic gain (RPG) in the active region. A high reflectivity (R > 99.5%) DBR stack made of 21-pairs of AlGaAs/AlAs is grown on the top of the active region. To avoid an immature thermal rollover, a detuning between quantum well gain peak and microcavity resonance of about 20 nm is introduced. This helps the VECSEL to have a robust performance at high temperature in expense of a slight increase in the threshold power.

Extraction of the waste heat from the active region is another challenge in the development of high power lasers. The VECSEL structure is designed for emission around 1178 nm with a barrier pumping at 808 nm. Such a large energy difference lowers the quantum efficiency of the laser and generates more waste heat in the active region. For efficient heat dissipation, after a thin Ti/Au metallization, the epitaxial side of the wafer is mounted on a CVD diamond using indium solder. The GaAs substrate is then completely removed by selective wet chemical etching. After substrate removal the remaining semiconductor is ~ 6 μm thick, allowing efficient heat dissipation at high pumping energy. The surface quality of the VECSEL sample is characterized by WYKO NT-2000 interferometer, and a peak to valley height of less than 40 nm over an area of 0.5 mm x 0.5 mm is measured. This optically smooth surface makes the scattering/diffraction loss negligible and results in high slope efficiency and high beam quality. Finally in order to reduce the surface reflectivity and sub-cavity resonance effect a dielectric anti-reflection layer is deposited. The processed VECSEL chip is mounted on a heat sink for temperature control.

EXPERIMENTAL RESULTS

To generate coherent fundamental and yellow-orange lasing, we used the folded cavity of figure 1, in which the VECSEL chip and a flat mirror serve as two end mirrors and a concaved spherical mirror as the folding mirror. The folding concave mirror with a ROC of 75 mm is high-reflective coated for the fundamental laser but is highly transmissive (~95%) at the yellow-orange wavelength. The full folding angle in the cavity is about 30 degrees. The laser is characterized by focusing an incoherent fiber-coupled 808-nm pump source on the chip with a spot size of 500 μm in diameter. In order to characterize the performance of the VECSEL in the ~ 1170 nm range a flat output mirror with a reflectivity of 96% is used. Linewidth narrowing and wavelength tuning is achieved by using an intracavity birefringent filter (BF), which is inserted at the Brewster angle. By using this low-loss filter, we can narrow the linewidth down to 0.5 nm and achieve over 30 nm wavelength tuning. Figure 2 shows the performance of the laser at 15 °C and 25 °C heat-sink temperatures for a 500 um pump spot. Maximum output power can reach 8.6 W at room temperature. The M^2 factor slowly increases from 1.03 at threshold to 1.5 at 8.6 W output, indicating a near TEM$_{00}$ transverse mode at high power operation.

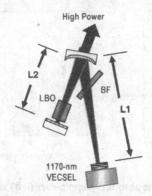

Figure 1. Schematic of the VECSEL setup for yellow-orange generation.

By replacing the 96% flat outcoupler with a highly reflecting mirror we build a high-Q cavity suitable for intra-cavity frequency doubling. In such a cavity the fundamental wavelength can be tuned from 1150 to 1195 nm. To generate yellow-orange lasing by frequency doubling, a LBO crystal with a size of 3 mm x 3 mm x 10 mm normally cut for the 1178 nm to 589 nm conversion is inserted close to the flat mirror, meeting a Type-I non-critical angular phase-matching condition. Both facets of the LBO are AR coated for both fundamental laser (1178 nm) and SHG signal (589 nm). In the tuning range, we lock the fundamental wavelength at 1159, 1170, 1178, and 1190 nm, respectively and align the phase matching angle φ of the LBO crystal to optimize the yellow-orange output. Figure 3-left shows the yellow output for a 500 um pump spot size. At this pump spot, over 5-W yellow-orange power at 585 nm and 589 is generated. For the 579 nm SHG optical to optical efficiency in excess of 12% is achieved. A wide tuning range covering 580 nm to 595 nm is also achieved (figure 3 right).

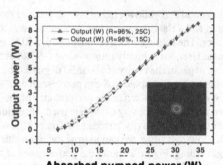

Figure 2. VECSEL fundamental output power vs. net pump power and the beam quality at 7 W output ($M^2 < 1.5$).

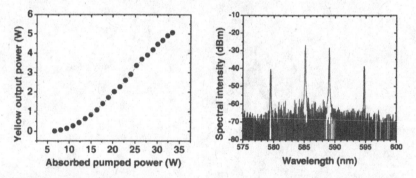

Figure 3. Yellow laser output power vs. the absorbed pump power (left) and yellow-orange wavelength tuning spectra (right).

CONCLUSION

We have demonstrated a highly strained InGaAs/GaAs tunable VECSEL laser with an output power in excess of 8 W at around 1173 nm. Full wavelength tuning in the 1150 - 1190 nm has been demonstrated using a folded high-Q cavity. High-power broadband output in the yellow-orange band (579.5-595 nm bands) is achieved through intracavity frequency doubling. Over 5-W yellow-orange tunable output power has been demonstrated over the 585-589 nm bands.

ACKNOWLEDGMENTS

The authors would like to acknowledge support from the U.S. Air Force Office of Scientific Research for a Phase II STTR grant Begin typing text here.

REFERENCES

1. N.S. Sadick and R. Weiss, J. Dermatol. Surg., **28**, 21-23 (2002).
2. C.F. Blodi, S.R. Russell, J.S. Padilo and J.C.Folk, Ophtalmology, **6**, 791-795 (1990).
3. C.E. Max, S.S. Oliver, H.W. Friedman, J. An, K. Avicola, B.W. Beeman, H.D. Bissinger, J.M. Brase, G.V. Erbert, D.T. Gavel, K. Kanz, M.C. Liu, B. Macintosh, K.P. Neeb, J. Patience, K.E. Waltjen, Science, **277**, 1649-1651 (1997).
4. Phillip A. Burns, Judith M. Dawes, Peter Dekker, James A. Piper, Jing Li, Jiyang Wang, Opt. Commun., **207**, 315-320, (2002).
5. D. Georgiew, V.P. Gapontsev, A.G. Dronov, M.Y. Vyatkin, A.B. Rulkov, S.V. Popov, J.R. Taylor, Opt. Express, **13**, 6772-6776 (2005).
6. Joshua C. Bienfang, Craig A. Denman, Brent W. Grime, Paul D. Hillman, Gerald T. Moore, and John M. Telle, Optics Letters, **28**, 2219-2221 (2003).
7. Evgeny Dianov, Alexey Shubin, Mikhail Melkumov, Oleg Medvedkov, and Igor Bufetov, J. Opt. Soc. Am. B, Doc. ID 75675 (posted 11/27/2006, in press).
8. L. Fan, M. Fallahi, J. T. Murray, R. Bedford, Y. Kaneda, J. Hader, A. R. Zakharian, J. V. Moloney, S. W. Koch and W. Stolz, Appl. Phys. Lett., **88**, 021105 (2006).
9. J. Hader, J.V. Moloney, M. Fallahi, L. Fan, S.W. Koch, Optics Lett., **31**, 3300 (2006).

Mater. Res. Soc. Symp. Proc. Vol. 1076 © 2008 Materials Research Society

Quantum Design of Active Semiconductor Materials for Targeted Wavelengths

Jerome Moloney[1,2], Joerg Hader[1,2], and Stephan W. Koch[3]

[1]Nonlinear Control Strategies, 3542 N Geronimo Ave, Tucson, AZ, 85705
[2]College of Optical Sciences, University of Arizona, 1630 E University Boulevard, Tucson, AZ, 85721
[3]Physics Department, University of Marburg, Renthof 5, Marburg, 35032, Germany

ABSTRACT

Performance metrics of every class of semiconductor amplifier or laser system depend critically on semiconductor QW optical properties such as photoluminescence (PL), gain and recombination losses (radiative and nonradiative). Current practice in amplifier or laser design assumes phenomenological parameterized models for these critical optical properties and has to rely on experimental measurement to extract model fit parameters. In this tutorial, I will present an overview of a powerful and sophisticated first-principles quantum design approach that allows one to extract these critical optical properties without relying on prior experimental measurement. It will be shown that an end device L-I characteristic can be predicted with the only input being intrinsic background losses, extracted from cut-back experiments. We will show that text book and literature models of semiconductor amplifiers and lasers are seriously flawed.

INTRODUCTION

Semiconductor amplifiers and lasers are pervasive as critical components in modern day technologies spanning low to high power applications[1]. The design and optimization of virtually every operational aspect of a semiconductor laser or amplifier requires a quantitative knowledge of the semiconductor material optical response. Important ingredients of the optical material properties are absorption/gain and refractive index, as well as radiative and nonradiative recombination processes. All of these quantities critically influence semiconductor amplifier or laser performance. Static properties such as threshold, slope efficiency, gain saturation, emission wavelength etc and dynamic properties including modulation response/ bandwidth, gain switching, mode-locking are cases in point. With very few exceptions, current semiconductor amplifier or laser design and modeling strategies involve a top-down approach whereby important macroscopic influences such as electrical and thermal transport within the complex semiconductor heterostructure are modeled with sophisticated packaged software tools[2-6]. The "Achilles Heel" of these approaches lies in the ad hoc manner in which they treat the semiconductor optical response - for the most part the above material properties are represented by phenomenological models that rely on external input which at best can be extracted from prior experimentally measured data.

Semiconductor wafer growth can now produce heterostructures of very high quality with stoichiometrically correct growth of individual mono-layers. Despite these significant advances in modern growth technologies, a critical void remains in predicting the performance of final packaged functional amplifier or laser devices. The lack of predictive semiconductor device design and growth monitoring capability can be traced to the extreme complexity of calculating the above semiconductor optical response from basic principles. The culmination of a series of important research breakthroughs over the past decade on the systematic calculation of semiconductor optical properties recently has led to the first ever prediction of an end-packaged semiconductor multiple quantum well (MQW) laser device performance without resorting to the use of adjustable fit parameters[7]. Now critical ingredients such as absorption/gain and refractive index (α-factor), spontaneous and Auger recombination rates can be computed for a wide range of material systems[8]. The systematic optimization of these systems and the potential cost savings of being able to fast-track to a final optimized semiconductor optical amplifier (SOA) or semiconductor laser (SL) structure for a targeted wavelength in very few, ideally a single wafer growth cycle, promises to have a wide economic impact across all semiconductor laser technologies. This has to be contrasted with current practice typically involving costly and time consuming multiple wafer growth/re-growth and packaging cycles before finalizing on a commercially feasible end product. At the end, one never knows whether the structure is truly optimal.

This paper will review the key breakthroughs made in the microscopic many-body approach to quantitatively calculating all of the above important ingredients of the semiconductor optical response of an active structure comprising one or more quantum wells. We will provide illustrative examples of a "bottom up" approach to designing final laser structures. Because our approach calculates the optical properties at the most fundamental level i.e the quantum wells, it applies to all types of laser structures independent of the device geometry. In order to place the results that follow in perspective, we reproduce the classical "rate equation" model for a semiconductor laser:

$$\frac{dS}{dt} = \Gamma G(N)(1 - \varepsilon S)S - \frac{S}{\tau_p} + \frac{\Gamma \beta_{sp} N}{\tau_N}$$

$$\frac{dN}{dt} = \frac{I}{qV} - \frac{N}{\tau_N} - G(N)(1 - \varepsilon S)S$$

Here the quantities S and N refer respectively to the photon and carrier densities in the active laser. The remaining parameters appearing in these equations are Γ the optical confinement factor, ε the nonlinear gain saturation parameter, τ_p the photon lifetime, β_{SP} the spontaneous emission factor, τ_N the carrier lifetime, q the electron charge, I the current and V the active volume. The gain G(N) is parameterized as $G(N) = G_0(N - N_{tr})$, where N_{tr} is the transparency density, while the carrier lifetime is expressed as $\tau_N = N / R(N)$ with R(N)=AN + BN^2 + CN^3. Built into this parameterization is the assumption that the laser is operating at a fixed wavelength (usually peak gain) and that recombination losses (at a fixed temperature) show the functional dependence shown for R(N) above. We will show that this assumed carrier dependence fails to describe any inverted semiconductor material below. This extensive parameterization is needed to fit experimentally measured data. The goal of the quantum design approach is to eliminate ad hoc parameter adjustment that requires prior experimental measurement and instead provide a predictive design approach that does not rely on experimental input. Of course, we cannot eliminate all parameters as some quantities such as defect recombination (AN term above), intrinsic material scattering and absorption losses are not amenable to quantitative calculation.

Another important empirical parameter in measuring laser performance is the so-called characteristic temperature T_0 that relates sensitivity of threshold current to temperature variation via the relation

$$J_{th} = J_0 \exp(T/T_0)$$

This simple relation, while having no obvious theoretical basis, has proved invaluable in determining performance of all classes of semiconductor lasers ranging from low-power edge and surface emitters to high power diode bars. A high value for T_0 indicates that the threshold current will be insensitive to temperature variation and consequently lead to optimal performance. Unfortunately, under current practice a device has to go through the full wafer growth and packaging cycle before this parameter can be determined from experimental measurements. The microscopic quantum design approach enables one to extract this parameter before any wafer is grown and moreover establish its dependence on fundamental microscopically calculated quantities such as spontaneous emission and Auger recombination losses. The temperature dependence of spontaneous and Auger losses also deviates strongly from that assumed in simple phenomenological models and this is discussed in the paper by Hader et al.[9] This paper connects T_0 to fundamental microscopic quantities and allows us to predict its value prior to wafer growth or laser packaging. Being able to compute the characteristic temperature prior to wafer growth will directly impact all semiconductor laser designs.

The closed-loop approach will generally entail designing the initial semiconductor epitaxial (epi) structure for a targeted wavelength, followed by wafer growth accuracy and quality verification and finally prediction of the end laser device performance. The next section begins with a brief overview of the basic many-body models that constitute the closed loop approach. These models have been validated by yielding quantitative agreement with experimental measurements of photoluminescence and gain spectra, spontaneous and Auger losses for a broad class of III-V and II-VI semiconductor material systems. As we do not start at an atomistic level, our approach requires accurate bandstructure parameters as input to the many-body calculations. These parameters are experimentally determined and well established for the GaAs/AlGaAS/InGaAs and InGaAsP/InP material systems. The latter material combinations provide for QW designs that approximately span the 800 nm – 1600 nm wavelength range. Less well established are the bandstructure parameters for material systems covering the visible-UV (GaN/InGaN) and mid-IR (GaSb/InGaSb) spectral regions. Despite this we have had significant success in applying our quantum design approach to these materials.

BANDSTRUCTURE AND MICROSCOPIC CALCULATIONS

As the central part for any predictive modeling of semiconductor laser systems one needs a reliable model for the gain materials[10,11]. Such a model should not only be able to describe the absorption and gain spectra as functions of carrier density and temperature, but also yield the refractive index, the luminescence and the nonradiative (Auger) losses. Ideally, the results of this model should be directly validated by detailed comparisons with quantitative experiments. The Semiconductor Bloch Equations (SBE) and Semiconductor Luminescence Equations (SLE) together with detailed bandstructure calculations, provide the theoretical foundation for extracting all relevant material optical properties for diverse material systems. Solution of this many-body problem is nontrivial but the effort in obtaining these solutions far outweighs the cost and effort invested in carrying out multiple trial and error growth and packaging cycles to determine a commercially successful laser designed for a targeted wavelength. As there are a number of many-body approaches discussed in the literature and implemented in a number of software packages, we briefly discuss the level of approximation needed to close the quantum design loop.

Bandstructure calculations carried out at the level of the $\vec{k} \bullet \vec{p}$ approximation generally suffice. This single electron theory can provide an accurate description of the relevant bands but cannot describe the dominant electron-hole many particle correlations within these bands. For this we need to solve the SBE model and include all relevant processes in a physically self-consistent manner. The simple two band version of the SBE equations that describe the induced polarization and inversion within and between all relevant sub-bands is given by[11,12]:

$$\left[i\hbar\frac{\partial}{\partial t} - \varepsilon_k^e - \varepsilon_k^h\right]P_k = \left[1 - f_k^e - f_k^h\right]\Omega_k + \frac{\partial}{\partial t}P_k\bigg|_{corr}$$

$$i\hbar\frac{\partial}{\partial t}f_k^a = -\Omega_k(t)P_k^* + \Omega_k^*(t)P_k + \frac{\partial}{\partial t}f_k^a\bigg|_{corr}$$

where $\Omega(t) = d_{cv}E(t) + \sum_{k'}V_{k-k'}P_{k'}(t)$ represents the generalized Rabi frequency or renormalized field and $\varepsilon_k^a = e_k^a - \sum_{k'}V_{k-k'}f_k^a$ are the renormalized energies. These renormalized quantities represent the only many-body effects that are taken into account in most theoretical treatments. The last terms in both equations above are often replaced by damping rates to represent irreversible decay to a heat bath by analogy with the classical optical Bloch equations. This assumption fails here as there are strong correlations within and between electron and hole plasmas in the various bands and one has to solve a quantum Boltzmann kinetic equation to properly account for these contributions. These calculations are much more involved than solving the reduced SBE equations themselves but are necessary to get quantitative agreement with the experimental semiconductor lineshape as discussed below. The electric field appearing in the SBE above is a classical quantity and at this level, we cannot account properly for spontaneous emission. To calculate the latter, we must treat the field itself as a quantum mechanical object and solve the Semiconductor Luminescence Equations. The details are not given here and the reader is referred to the literature for details[13].

With these many-body models we have a quantum-mechanical theory for the electron-hole system in the active semiconductor material. In order to base the calculations on as little empirical input as possible, we evaluate in a first step the relevant part of the material's bandstructure. This can be done at the level of the well-established k.p theory where only few input quantities, such as the Luttinger parameters and the band-offsets are needed[14-16,10]. In many cases, these inputs are known from independent investigations, allowing one to have a high level of confidence in the resulting bandstructure calculations. These calculations not only yield a detailed description of the relevant valence and conduction bands of the gain material, but one also obtains the corresponding electronic wavefunctions needed e.g. to evaluate the transition-dipole matrix elements that determine the relative strengths of the optically allowed interband transitions. An example of such a computed bandstructure for InGaPAs is shown in figure 1.

Despite the great success of these bandstructure calculations for a wide range of material systems, one has to remember, however, that often for less well-studied and novel materials important input parameters may not be known with high precision. For example, in a quantum-well heterostructure, the alignment between the conduction-band minima and valence-band maxima of quantum-well and barrier material can be uncertain. Sometimes the possibilities range even from the standard type-I alignment, where the conduction-band minimum is lower and the valence-band maximum is higher in the quantum well than in the barrier, to a type-II alignment where the conduction-band minimum may be higher in the well than in the barrier. In such a situation, one needs feedback from independent experiments to reduce the level of uncertainty.

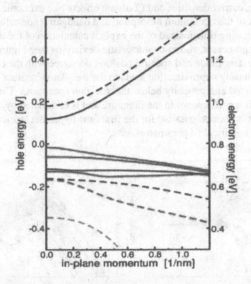

Figure 1. Computed bandstructure for a 6nm wide $In_{0.9}Ga_{0.1}As_{0.53}P_{0.47}$ well between 10nm wide $In_{0.88}Ga_{0.12}As_{0.26}P_{0.74}$ barriers and InP cladding layers. Black: Electron-bands. Red: Heavy hole bands. Blue: Light hole bands; Green: Split-off hole bands. Solid: Subbands confined in the well; Dashed: Bulk bands of the barrier material.

After the bandstructure and the matrix elements are known, the next important step is to compute the optical material response. For this purpose, it is crucially important to correctly deal with the Coulomb interactions in the system of charge carriers. Furthermore, one has to keep in mind that electrons and holes are Fermionic quasi-particles, for which the Pauli exclusion principle states that each quantum state can be occupied at most by a single particle. This Pauli exclusion principle and the other many-body interaction effects are correctly included in the Semiconductor Bloch Equations[10,11], which in a multi-band generalized form are by now the basis for most microscopic gain theories.

For all optical systems, and in particular for semiconductor lasers it is crucial to properly model the phase destroying processes since these are responsible for the loss of optical coherence and for the spectral lineshapes. Based on the fundamentally new concept of excitation induced dephasing (EID)[17,11] it became possible for the first time to systematically compute the temporal decay of the optical polarization for the relevant densities and excitation conditions. The EID concept replaces the old dogma of constant dephasing times with a microscopic theory that allows us to compute the density-, temperature-, and frequency-dependent decoherence rates.

The microscopically correct description of the interaction and dephasing processes lead to significant deviations from the Lorentzian lineshape that is well known from atomic systems. As one can see in the example in Figure 2 (Left) the fully microscopic theory (solid lines) agrees very well with the experimental results (dots). On the other hand, the identical calculation, using the same bandstructure, carrier densities and Coulomb effects like excitonic resonances, Coulomb enhancement of the continuum absorption and bandgap renormalisation, but using a phenomenological dephasing time instead of the explicit calculation of the underlying microscopic scattering processes, shows characteristic deviations (see Figure 2 Right). Not only does the gain lineshape, amplitude and spectral position not agree with the experiment, but the phenomenological dephasing approximation leads to the well-know artifact that absorption (negative gain) is predicted energetically below the true gain spectrum. This semiconductor-laser lineshape problem is well known in the literature and is eliminated by the microscopic approach. On this basis it became possible for the first time to predict semiconductor laser gain spectra without ad hoc empirical fit parameters.

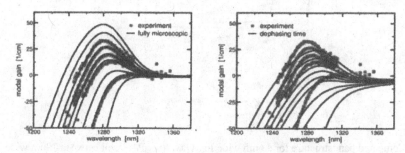

Figure 2. Left: Comparison of microscopically computed and experimentally measured gain spectra for an InGaPAs QW as a function of increasing carrier density (bottom to top curves). **Right:** Same experimental data but using phenomenological dephasing times.

Wafer-Level Diagnostics: From Photoluminescence to Gain

After the epitaxial growth of the semiconductor wafer, it is desirable to perform quantitative tests before the structure is moved on to further processing steps. In particular it would be helpful to know already at this wafer level, if and to which degree the amplification and gain characteristics of the structure satisfy the specified requirements. For this purpose, it is useful to employ photoluminescence measurements taken at relatively low excitation intensities which corresponds to low internal carrier densities in the excited semiconductor structure. Because of the low density, the luminescence does not provide direct information about the preferred wavelength at which the strongly pumped, inverted semiconductor material will lase. Here, the usual recipe is to assume that the laser wavelength will generally be shifted 6-10 nm down from the low density PL peak! However, such guesses are uncertain at best, often they are incorrect. To overcome this problem, we propose to use the fully microscopic theory that straightforwardly allows us to extract the high density material gain from the same calculation as the low density luminescence. The first validation of the connection between low carrier density PL and high density inverted semiconductor gain was made possible via collaboration between a semiconductor growth team at AFRL, the University of Arizona ACMS research group and an experimental group at the University of Bochum in Germany in 2002[18]. The sequence of steps involved in this validation procedure provide the first demonstration of a closed-loop wafer level diagnostic coupled to gain measurement on the processed sample.

Step 1. A structure with three InGaAs wells between GaAs barriers was grown by MBE. The measured photoluminescence spectra were provided at five excitation pump levels together with the nominal growth parameters i.e the well widths of 5 nm and Indium concentrations of 20%. The measured PL spectra are shown in blue in Figure 3.

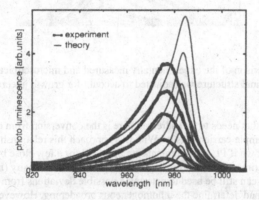

Figure 3. Experimentally measured photoluminescence spectra (blue curves) at different low level illumination intensities and corresponding microscopically computed PL spectra for a series of increasing carrier densities. The latter were computed for the nominal $In_{20}Ga_{80}As$ QW structure believed to have been grown by the experimentalists.

Step 2. The photoluminescence spectra were then computed for the ideal semiconductor crystal using the microscopic theory and the nominal growth parameters supplied. The corresponding PL spectra are shown in red in Figure 4. There is a clear misalignment between the theory and calculated PL spectra and, moreover, the theoretical spectra are somewhat narrower. The 6 nm shift in PL spectral peaks can be accounted for by changing the In concentration from the nominal value of 20% to about 19% or reducing the well width by about 1nm. As the former is more likely, the Indium concentration was adjusted to 19% and the PL spectra re-calculated. Additionally, inevitable growth fluctuations during the MBE deposition lead to an inhomogeneous broadening of the PL peaks. By applying an inhomogeneous broadening of 16meV (FWHM), we obtained the PL spectra shown in Figure 4. The PL peaks are perfectly aligned and the relatively small degree of inhomogeneous broadening indicates a very high quality MBE growth. We emphasize that the Indium concentration adjustment and inhomogeneous spectral broadening are related to growth and not, theory uncertainties.

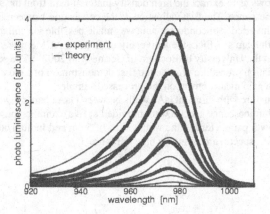

Figure 4. New comparison of the experimentally measured and microscopically calculated PL spectra when the nominal structure was adjusted to account for growth uncertainty.

Another issue that needs to be addressed here is the conversion from external pump laser intensity to internal sample carrier density. With this approach this relationship can be unambiguously tied down if PL measurements are provided for a few close by illumination intensities. If the experimental PL is known only for one excitation density (in the low excitation regime), the approach can still be used to determine possible deviations from the nominal structural parameters and determine the inhomogeneous broadening. However the exact correspondence between excitation power and intrinsic carrier density can no longer be established. This is due to the fact that in the limit of low excitation densities the PL-lineshape and spectral position become excitation independent and only the amplitude changes. It should be noted however, that in the experiment the excitation density has to be chosen high enough such that the measured PL is not dominated by the emission from defect states below the bandgap.

Step 3. Now that we have verified the actual growth structure, we simply compute the inverted material gain for this structure. Importantly, it is vital to compute the gain for the homogeneously broadened epi structure and apply the calculated inhomogeneous broadening to the calculated gain spectra. To test the level of accuracy of the theoretical predictions, an experimental measurement of the gain from the exact same wafer was performed. The experimental gain curves (blue with dots) were found to lie on top of the theoretically computed gain spectra as shown in Figure 5. This to our knowledge was the first time that an experimental measurement verified a prior theoretical calculation!

The above sequence of steps provides a systematic predictive and adjustable-parameter-free methodology for characterization semiconductor wafer growth.

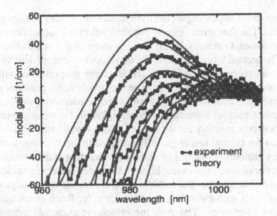

Figure 5. Experimentally measured gain spectra agree with the calculated spectra.

RADIATIVE AND NONRADIATIVE LOSSES

The microscopically computed photoluminescence and gain spectra discussed in section two provide the critical foundation for designing a semiconductor active region for a targeted wavelength. At this level, they provide invaluable wafer analysis tools and direct feedback to the semiconductor laser designer and wafer grower on the accuracy and quality of the actual grown wafer. However to move to the next step and design a final semiconductor laser device we need to know the sources of current losses in the device[19-23,1]. The widely accepted model for such losses is the well-known "ABC" whereby carrier density dependent current losses at a fixed temperature are represented by the formula:

$$J_{Loss} = AN + BN^2 + CN^3$$

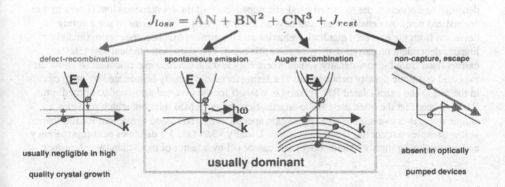

$$J_{loss} = AN + BN^2 + CN^3 + J_{rest}$$

defect-recombination   spontaneous emission   Auger recombination   non-capture, escape

usually negligible in high quality crystal growth   **usually dominant**   absent in optically pumped devices

The individual terms appearing on the right hand side of this equation have the following physical interpretation. The first term, linearly proportional to the carrier density N, accounts for defect recombination losses. Generally in high quality material growth, this term is not a significant player. The second term, quadratically dependent on carrier density N, accounts to radiative or spontaneous recombination losses. The last term, proportional to the carrier density cubed, accounts for nonradiative Auger losses. The above schematic provides a pictorial representation of these individual loss terms as well as some additional sources of loss.

The 'non-capture, escape' losses are even negligible in electrically-pumped semiconductor lasers except possibly for the case of shallow wells with weak carrier confinement. This law for current loss is globally accepted in the literature and the assumptions contained in it, have a profound influence of almost all operating characteristics of a semiconductor amplifier or laser. For example, such loss rates determine static properties such as laser threshold, slope efficiency, thermal rollover etc and intrinsic dynamic properties such as modulation rate/bandwidth, gain switching, mode-locking, feedback instabilities, intensity filaments in broad area emitters etc. The usual prescription in using this formula in laser modeling is to extract the "A,B,C" coefficients from fits to experimental measurements on known semiconductor samples. Usually, such measurements are carried out under low excitation conditions where the carrier (electrons and holes) approximately obey Boltzmann statistics. Moreover, these "ABC" coefficient values are then taken as fixed parameters for the semiconductor material system in question and are then applied to other lasers in the same material class.

Microscopic calculations carried out at the same level of sophistication as the gain measurements, have demonstrated unequivocally that the empirical ABC formula becomes invalid for inverted semiconductor laser media[24]. This should not be too surprising given the fact that for high densities the Boltzmann approximation for the carrier distributions becomes invalid and one has to use the correct Fermi-Dirac statistics. However, the situation is even more complex however since the loss mechanisms additionally exhibit sensitive dependencies on the many-body interactions in the electron-hole plasma, on the QW widths etc. The end result is that the carrier dependencies assumed for spontaneous and Auger losses differ dramatically from the assumed dependence in the "ABC" law.

Figure 6 shows plots of the spontaneous current loss as a function of carrier density for three temperatures in a 6.4 nm GaInNAs quantum well laser emitting at 1.3μm. At low carrier densities one observes the expected quadratic dependence of the spontaneous loss (linear in the normalized plot). At densities near and beyond the transparency one begins to see a strong deviation from the expected quadratic behavior and the current loss becomes approximately linearly dependent on carrier density beyond this point. The phenomenological loss rates are extrapolated from the low-density values where they and the microscopic calculations show the expected quadratic density dependence. The lasing threshold density is indicated by open circles in the plots. The extrapolated BN^2 relation is way off for the inverted semiconductor medium. We note here that the often used Kubo-Martin-Schwinger (KMS) relation, which provides a simple integral-conversion of absorption/gain spectra, only gives good agreement for lineshapes at low densities but not too low temperatures. Usually KMS fails for densities near transparency and above (lasing threshold and above) and can be off by a factor of more than two. A proper microscopic calculation of spontaneous losses additionally requires quantization of the light field which then leads us to the evaluation of the Semiconductor Luminescence Equations[13,25].

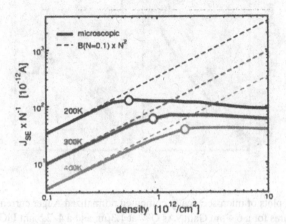

Figure 6. Log-log plot of the microscopically compute spontaneous current losses scaled inversely with carrier density as a function of carrier density at three temperature. The normalization to carrier density ensures that the spontaneous current loss term in the "ABC" law is a linear function of carrier density as shown. The phenomenological BN^2 dependence is shown as a series of straight lines.

The microscopic calculation of Auger processes is even more involved and has only recently been achieved. Details of the calculations are provided in reference 24. Here we present microscopic calculations of both spontaneous and Auger processes for three different material systems and show that these agree quantitatively with measured data, again without using adjustable fit-parameters.

Figure 7 shows log-log plots of the Auger loss rates for two material systems, a 1.3μm 6.4nm GaInNAs QW and a 1.5μm 4×2.5 nm InGaPAs material. The calculations are shown for three different temperatures. The phenomenological loss rates are again extrapolated from the microscopic low-density values. Strong deviations from the microscopically computed loss rates are again evident at and beyond transparency density. The data in figures 6-7 show the density dependent radiative spontaneous and non-radiative Auger loss rates at a given temperature. The experimentally measured temperature dependence[26] of these rates is depicted in figure 8 together with microscopic calculations for a 1.3μm and a 1.5μm InGaPAs laser.

The theory-experiment comparison in figures 7 shows that the combined spontaneous and Auger currents agree quantitatively with the measured data over the broad temperature range indicated. The slight deviations at high temperature are probably due to internal heating of the device beyond the heat sink temperature. Quantitative agreement of the spontaneous and Auger loss currents over a similar temperature range were also with experimental measurements for a 6.4nm GaInNAs QW structure lasing at 1.3μm.

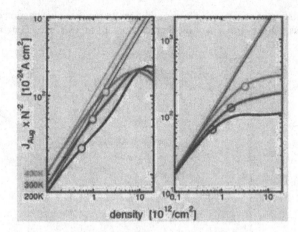

Figure 7. Log-log plots of microscopically computed normalized Auger current losses at three different temperatures for a 6.4 nm GaInNAs QW at 1.3μm and a 4×2.5 nm InGaPAs QW at 1.5μm. The Auger loss rates are divided by N^2 to yield a linear plot for the phenomenological "CN^3" loss rates.

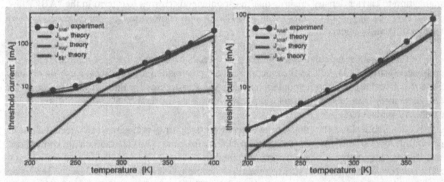

Figure 8. Comparison of experimentally measured and microscopically computed spontaneous and Auger threshold currents as a function of temperature for a 1.3μm and a 1.5μm InGaPAs QW laser.

DESIGNING HIGH POWER VERTICAL EXTERNAL CAVITY SEMICONDUCTOR LASERS

Optically-pumped vertical external cavity surface emitting lasers (VECSELs) or equivalently, optically pumped semiconductor lasers (OPSLs) provide a new class of high power, high brightness semiconductor laser sources. In contrast to broad area devices where the output beam suffers strong astigmatism due to differing beam divergence along the fast and slow axis, the VECSEL output can be confined to a clean TEM$_{00}$ beam. Moreover, broad area emitters undergo strong dynamic filamentation along the slow axis leading to broad far-field outputs well in excess of the expected far-field divergence of a stable beam. The VECSEL is akin to a mini-disk-solid state laser but with many advantages over the latter. Solid-state mini-disk lasers typically require narrow wavelength pumps and multiple passes back and forth through the structure to achieve sufficient pump absorption. Barrier pumped VECSELs can absorb all of the pump light in either a single or double pass through an extremely thin (5 – 10 μm) active layer and can use off-the-shelf low cost pumps requiring no temperature stabilization. Because VECSELs are based on semiconductor QW materials, they provide wavelength agile systems and have already been demonstrated to operate around 675nm (InGaP)[31] , 850nm (GaAs)[30], 980 nm (InGaAs)[27,28,29], 1500 nm (InGaPAs)[32] and 2400nm (GaSb)[33].

The key design principles of a VECSEL structure resemble that of the more common low-power vertical cavity surface emitting lasers (VCSELs). The key difference with VCSELs lies in the fact that the top high reflectivity DBR mirror is removed and replaced by a lower reflectivity external cavity curved mirror. The external cavity mirror reflectivity can range from 90% to 98% requiring the addition of many QWs to provide enough single pass gain to offset these relatively large losses. The lower external mirror loss facilitates greater power extraction from the optically pumped spot on the VECSEL chip and power scaling is achieved by increasing the pumped area on the chip. With this particular geometry, Coherent Inc of Santa Clara, California[34] have demonstrated up to 50 Watts output from a single 900μm pump spot at 1050 nm. A schematic of the VECSEL structure is shown in Figure 9.

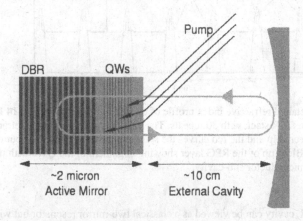

Figure 9. Schematic of a vertical external cavity semiconductor laser. The active semiconductor chip consists of a MQW resonant-periodic gain stack grown on a high reflectivity DBR mirror. Typically the structure is optimized with 10-14 QWs in the RPG stack. The active chip is mounted on a high thermal conductivity CVD diamond heat spreader and the latter is typically mounted on a copper heat sink. The incoherent pump light from the diode bars is incident at set to a 45 ° angle and an external curved reflector at a distance of between 10-20 cm provides external feedback of light to the active semiconductor mirror.

The active semiconductor chip consists of a resonant periodic gain (RPG) structure consisting of multiple QWs arranged to coincide with the anti-nodes of the standing wave optical field within the semiconductor sub-cavity. The QWs are grown on a high reflectivity DBR stack containing on the order of 20-30 repeats of, for example, AlGaAs/AlAs pairs (see Figure 10). The active chip consisting of the RPG and DBR stack is mounted on a high thermal conductivity heat spreader such as CVD diamond so as to efficiently extract the heat from the hot active RPG layer. The VECSEL cavity is completed by adding an external reflector typically places around 10-20 cm from the active chip as shown in Figure 9. Finally the pump light (at 808 nm for example) is incident on the active chip at an angle close to 45°.

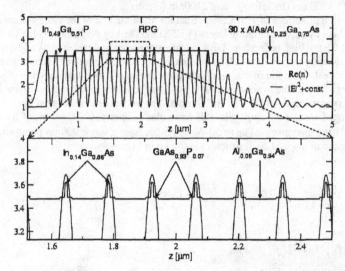

Figure 10. **Top:** Detailed refractive index profile of the active RPG region with 14 InGaAs QWs and the passive DBR stack with 30 repeats. The blue curves represent the refractive index along the active mirror chip and the red curves the standing wave intensity distribution along the structure. **Bottom:** Blow-up of the RPG layer showing alignment of the QWs with the standing wave field in the semiconductor sub-cavity.

The VECSEL cavity can be viewed as a classical two-mirror resonator but with the important proviso that the reflectivity of one mirror (the active semiconductor chip) can be actively controlled via the external diode bar pumps[35,36]. The reflectivity of this active mirror as a function of coupled pump power is shown in Figure 11. At low external pump powers (internal sheet carrier density of 0.05×10^{12} cm^{-2} – solid black curve in the figure) the active chip reflectivity spectrum shows a pronounced absorption dip within its photonic stop-band due to strong absorption of the pump light by the QWs. The absorption of the QW stack is pre-computed using our many-body microscopic approach and represents a critical input to the VECSEL design. Increasing the pump power to the point where the QWs are inverted (internal sheet carrier density of 3.5×10^{12} cm^{-2} – solid green curve in the figure) leads to a narrow gain peak (reflectivity > 1) superimposed on the sub-cavity stop-band. Note that the gain peak is shifted to longer wavelengths relative to the absorption dip due to the internal heating of the chip (now 360°K versus the cold-cavity 300°K temperature at low pump levels). The strong shift in material gain peak with increasing pump intensity (carrier density and temperature) must be allowed for in the VECSEL cavity design. Additionally the internal active chip resonance due to the DBR stack and semiconductor – air interface shifts to longer wavelength with increasing temperature but this shift is much less than the gain peak shift. Our cold-cavity design takes accounts of these relative shifts and optimizes the structure to avoid early shut-off due to thermal rollover.

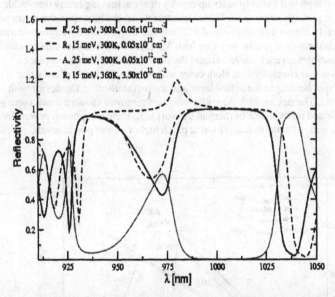

Figure 11. Reflectivity (R) and absorption (A) spectra of the active semiconductor mirror as a function of incident pump power. Low pump levels show an absorption dip in the photonic stop-band at the central resonance of the MQW RPG stack (solid black and dashed red curves). The corresponding absorption spectra are shown in green for the same conditions. At higher pump levels (carrier sheet density of 3.5×10^{12} cm^{-2}) and at the elevated temperature of the running "hot" cavity (blue dashed curve) the mirror exhibits a narrow gain peak superimposed on the stop-band.

Designing a high power VECSEL structure optimized to extract maximum power at, say 975 nm when pumped by an 808 nm diode bar, requires optimization at many levels. The underlying semiconductor active sub-cavity has to be designed to exhibit the maximum gain at the internal temperature of the running laser. The latter depends sensitively on thermal management. As we mentioned earlier, the gain peak can shift rapidly as a function of increasing temperature in the active layer. Typically an internal temperature elevation relative to ambient of about 100°K can be tolerated before the device shuts off. Efficient heat removal from the active layer is the key to maximizing the power extraction – the latter depends on a number of factors including high thermal conductivity heat spreaders as close as possible to the active RPG layer. The most effective design to date has involved growing the VECSEL chip (DBR+RPG) on a substrate with the DBR layer on top. In this setup, the chip is typically mounted on a CVD diamond heat spreader and the entire GaAs substrate has to be removed by chemical etching. These processing steps are extremely critical as the end active mirror is only 6-10μms thick. If the chip is grown in the conventional manner, the thermal impedance through the DBR and GaAs substrate is too high and the chip heats up rapidly never achieving lasing threshold. The only option here is to partially remove the GaAs substrate to get the heat spreader closer to the active region. Figure 12 shows a comparison of the situation above with the entire substrate removed against a conventional grown structure with 5μm and 7.5μm of substrate remaining after etching. The numbers on each curve indicate the internal temperature along each characteristic input-output characteristic. Both cases with a finite thickness of substrate remaining show a rapid heating in the active layer and a sharp shut-off in the device with increasing pump level. The device with the entire substrate removed shows a much more gradual temperature increase and no evidence of thermal shut-off with increasing pump power. In practice, this device will ultimately shut-off but at much higher output power levels.

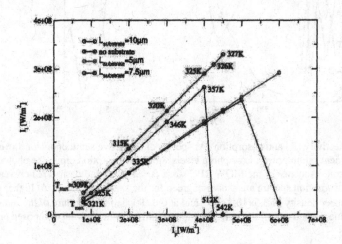

Figure 12. Input power density versus output power density for four different VECSEL structures (see captions).

An interesting optimization strategy is to remove the heat directly from the top of the chip by capillary bonding a transparent single crystal diamond heat spreader on top and actively water cooling the intracavity heat spreader. This approach has been shown to work for conventionally grown VECSELs by providing a much lower thermal impedance pathway via the high thermal conductivity diamond. However, extracting the heat simultaneously from both top and bottom of the chip offers even greater power scaling possibilities. Figure 13 contrasts the extracted output power for three cases with different cooling setups. In each case the pump spot is 400 μm. The leftmost black curve corresponds to the power extracted from our first device when the heat extraction is through the DBR to the CVD diamond heat spreader i.e no top heat spreader. By adding a transparent diamond heat spreader to the top of this device, a dramatic improvement in extracted power is predicted as shown by the red curve. Even keeping the conventional device with 50 μm of GaAs remaining but with a top and bottom heat spreader gives significantly improved extracted power as shown by the green curve.

These VECSEL structures offer enormous flexibility as regards wavelength selectivity, tunability, high power high brightness operation, wavelength generation in the visible through intra-cavity SHG (red at the fundamental, blue, green and yellow at the second harmonic have been generated). The near TEM_{00} fundamental mode operation can be achieved by carefully matching the incident pump spot to the laser signal spot. The next section recaps some experimental results from the University of Arizona where the VECSEL chips were designed prior to wafer growth.

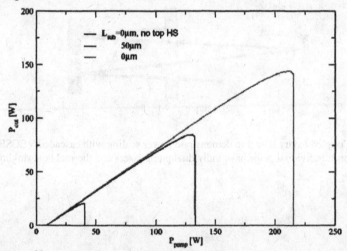

Figure 13. Input vs output power for a VECSEL chip with a 400μm pump spot diameter. The lower left black curve corresponds to the case where the substrate is completely removed and the device in mounted on a CVD diamond heat spreader. The green curve is when additionally a transparent intracavity diamond heat spreader is capillary bonded to the top RPG stack end of the device. The red curve is a conventional structure with 50μm of the GaAs substrate remaining for mechanical support and an additional intracavity transparent diamond heat spreader boned to the top of the RPG stack.

MULTI-WATT TUNABLE VISIBLE VECSEL LIGHT SOURCES

In this section, we briefly review some key applications of VECSEL structures designed by the above quantum closed-loop approach. The two-mirror cavity (one active and one passive) is ideal for spectral selectivity, wavelength tunability and visible light generation via intra-cavity SHG. By placing a birefringent filter in the cavity, it is possible to spectrally narrow and tune the laser output over 30-50nm range.

Figure 14 shows a folded cavity layout to demonstrate power scaling of individual VECSEL chips designed to lase at 975nm. Each chip is separately pumped and has separate heat sinks. This set-up allows for enhanced wavelength tunability relative to a singe chip. Figure 15 compares the input-output power characteristics of VECSEL cavities with the individual chips to a single cavity arrangement with both chips as active mirrors. We observe that the output power is scaled up by a factor of two with no reduction in slope efficiency. Additionally, the wavelength can be tuned over a 30nm range by adjusting a birefringent filter in the cavity. The output power of these chips is limited by the available pump spot area of less than 500 μm – the latter is limited by surface defects that act as scattering losses. We anticipate that new wafer growths with much larger pumped surface areas will yield very significant power scaling up to hundreds of Watts and beyond.

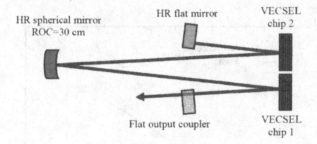

Figure 14. Coupled cavity layout to demonstrate power scaling with cascaded VECSEL chips (active mirrors). Individual chips have individual pump lasers and thermal heat sinking.

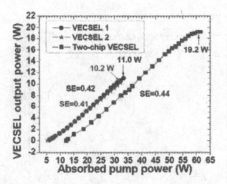

Figure 15. Input output power characteristics for individual and a cascaded VECSEL cavity.

Our second example is of a VECSEL chip designed to lase at a fundamental wavelength of 1178nm with the goal of generating a multi-Watt yellow output at 589nm. The 10 QW RPG stack consisted of highly strained InGaAs QWs with GaAsP strain compensating barriers. The QW gain for these structures is higher than that at 975nm. The InGaAs MQW epi design was grown using low temperature MOVPE. The MOVPE growth uses novel alternative liquid MO-V-sources (TBAs, TBP) that decompose at lower temperatures than conventional hydride precursors. This allows for a general reduction of the growth temperatures, which in turn makes it possible to realize structures with higher values of strain and, thus, higher In-concentrations reproducibly in the active QW. In addition, also the realization of the necessary strain-compensating GaAsP barrier layers with precise chemical composition is facilitated. This enabled us to push the InGaAs epitaxial structure towards the target wavelength of 1178 nm. The RPG structure itself consisted of a 10-repeat, 7 nm InGaAs QW with GaAsP strain compensation barriers, separated by AlGaAs barriers. A folded cavity four-mirror arrangement is used to extract the SHG signal. The VECSEL chip and a highly reflective flat mirror act as the end mirrors. The highly reflective concaved mirror (with 75 mm radius of curvature) acts as a folding mirror.

A type-I phase matched LBO crystal (3 mm X 3mm X10 mm) was aligned close to the flat mirror. The diameter of the pump spot is about 500µm. The mirror geometry has been employed to generate up to 5 Watts of yellow light in a TEM$_{00}$ mode at 589 nm. The input-output power of the second harmonic at 589 nm is shown together with the fundamental in Figure 16. The birefringent filter is used to tune the cavity to this wavelength and the output is spectrally narrow but not single mode. Single mode emission can be achieved by inserting a Fabry-Perot etalon in the cavity. Our next goal is to optimize the extracted signal power and reduce the linewidth further.

The wavelength tunability of this chip at the fundamental wavelength ranged from 1147nm to 1197nm.

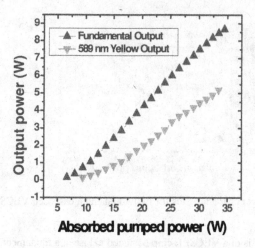

Figure 16. Input-output power characteristic of the fundamental at 1178nm and the second harmonic signal at 589nm.

CONCLUSIONS

We have discussed and demonstrated a first principles quantum design approach that removes ad hoc parameterization from semiconductor laser design. The approach is very general requiring only measured bandstructure parameters and band offsets as input. In this sense, III-V and II-VI semiconductor material systems yielding lasing output from the visible/UV to mid-IR can be treated on the same footing. The approach is being currently applied to GaN/InGaN QWs for UV/visible emission and to InGaSb/GaSb materials to cover the important mid-IR band[42,] VECSELs have already been demonstrated in the 2.3-2.4 μm wavelength window and a challenge is to push these to longer wavelengths covering the 3-5μm band and beyond. Thermal management will be a major challenge at longer wavelength as Auger processes tend to dominate. Currently devices at these wavelengths tend to operate as edge emitters often at cryogenic temperatures. VECSELs could offer an attractive alternative to QCL and ICL lasers currently under development.

REFERENCES

1. G.A. Agrawal and N.K. Dutta, Long-Wavelength Semiconductor Lasers, 2nd Ed., Van Nostrand Reinhold Co., New York (1986).
2. K. Hess, Advanced Theory of Semiconductor Devices, IEEE Press, New York (2000).
3. Crosslight software http://www.crosslight.com .
4. Rsoft software http://www.rsoftdesign.com .

5. J. Piprek, Semiconductor Optoelectronic Devices, Academic Press, San Diego (2003).
6. J. Piprek (Ed.), Optoelectronic Devices, Springer Verlag (2005).
7. J. Hader et al., Opt. Letts. **31**, 3300 (2006).
8. J.V. Moloney, J. Hader and S.W. Koch, Lasers & Photonics Reviews, **1**, 24 (2007).
9. J. Hader, J.V. Moloney and S.W Koch, "Temperature dependence of radiative and Auger losses in quantum well lasers", Conference 6889 Physics and Simulation of Optoelectronic Devices Paper 6889-09.
10. W. Chow and S.W. Koch, *Semiconductor Laser Fundamentals,* [Berlin, Heidelberg: Springer-Verlag] (1999).
11. H. Haug and S.W. Koch, Quantum Theory of the Optical and Electronic Properties of Semiconductors, 4th Ed., World Scientific Singapore (2004).
12. M. Lindberg and S.W. Koch, Phys. Rev. B 38, 3342 (1988).
13. M. Kira and S.W. Koch, Prog. Quant. Electron., in print (2007).
14. J. Callaway, Quantum Theory of the Solid State, Part A, Academic Press, New York (1974).
15. M. Altarelli, p.12 in Heterojunctions and Semiconductor Superlattices, Eds. G. Allan et al, Springer Verlag, Berlin (1985).
16. G. Bastard, Wave Mechanics Applied to Semiconductor Heterostructures, Les Editiones des Physiques, Paris (1988).
17. S. Hughes et al., Solid State Comm. 100, 555 (1999).
18. J. Hader et al., *IEEE Phot. Tech. Letts.*, **14**, p762 (2002).
19. A. Yariv, Quantum Electronics, 2nd Ed., Wiley, New York (1975).
20. G.H.B. Thompson, Physics of Semiconductor Lasers, Wiley, New York (1980).
21. P.S. Zory, Quantum Well Lasers, Academic Press, San Diego (1993).
22. L.A. Coldren and S.W. Corzine, Diode Lasers and Photonic Integrated Circuits, Wiley, New York (1995).
23. S.L. Chuang, Physics of Optoelectronic Devices, Wiley, New York (1995).
24. J. Hader, J.V. Moloney, S.W. Koch, IEEE J. Quantum Electron. **41** No. 10 (2005).
25. M. Kira et al., Prog. Quant. Electron. 23, 189 (1999).
26. A.F. Phillips et al., IEEE J. Sel. Top. Quant. Electron., **5**, 401 (1999).
27. M. Kuznetsov et al., IEEE J. Sel. Topics Quantum Electron., **5**, 561 (1999).
28. J. Chilla et al, Proc. SPIE Int. Soc. Optical Engineering, **5332**, 143 (2004); Proc. SPIE **6451**, 645109 (2007).
29. S. Lutgen et al., Appl. Phys. Lett., **82**, 3620 (2003).
30. J.E. Hastie et al., IEEE Phot. Tech. Letts., **15**, 894 (2003).
31. J.E Hastie et al., Opt. Express, **13**, 77 (2005).
32. H. Lindberg et al., IEEE Phot. Tech. Letts., **16**, 362 (2004).
33. N. Schulz et al., IEEE Phot. Tech. Letts., **18**, 1070 (2006).
34. J. Chilla et al., Proc. SPIE (2001).
35. A.R. Zakharian et al., Appl. Phys. Lett., **83**, pp. 1313-1315 (2003).
36. A.R. Zakharian et al., IEEE Photon. Technol. Lett. **17**, 2511-2513 (2005).
37. Li Fan et al., Applied Physics Letters **88**, pp.021105 (2006).
38. Yushi Kaneda, et al., IEEE Photonics Technology Letters, **18**. 1795 (2006).
39. Li Fan et al., Applied Physics Letters, **88**, 251117 (2006).
40. Li Fan et al., Appl. Phys. Lett., **90**, 181124, (2007).
41. Li Fan et al., Appl. Phys. Lett. **91**, 131114 (2007).
42. G. Balakrishnan et al., Lase 2008 Solid State Lasers XVII, Paper 6871-34.

Mater. Res. Soc. Symp. Proc. Vol. 1076 © 2008 Materials Research Society　　　　　1076-K07-06

Advanced Laser Diode Cooling Concepts

Ryan Feeler[1], Jeremy Junghans[1], Edward Stephens[1], Greg Kemner[1], Fred Barlow[2], Jared Wood[2], and Aicha Elshabini[2]

[1]Cutting Edge Optronics, 20 Point West Boulevard, St. Charles, MO, 63301
[2]Electrical Engineering, University of Idaho, Buchanan Engineering, Room 213, PO Box 441023, Moscow, ID, 83844-1023

ABSTRACT

A new, patent-pending method of cooling high-power laser diode arrays has been developed which leverages advances in several areas of materials science and manufacturing. This method utilizes multi-layer ceramic microchannel coolers with small (100's of microns) integral water channels to cool the laser diode bar. This approach is similar to the current state-of-the-art method of cooling laser diode bars with copper microchannel coolers. However, the multi-layer ceramic coolers offer many advantages over the copper coolers, including reliability and manufacturing flexibility. The ceramic coolers do not require the use of deionized water as is mandatory of high-thermal-performance copper coolers.

Experimental and modeled data is presented that demonstrates thermal performance equal to or better than copper microchannel coolers that are commercially available. Results of long-term, high-flow tests are also presented to demonstrate the resistance of the ceramic coolers to erosion. The materials selected for these coolers allow for the laser diode bars to be mounted using eutectic AuSn solder. This approach allows for maximum solder bond integrity over the life of the part.

INTRODUCTION

Recent advances in semiconductor technology have led to the creation of laser diode bars capable of producing hundreds of watts of CW output power. These devices typically operate with electrical-to-optical efficiencies in the range of 50-75%. As a result a tremendous amount of waste heat is generated, with heat fluxes on the order of 1 kW/cm^2 common in the industry today. As device technology continues to improve and optical output powers continue to increase, additional waste heat will need to be removed by laser diode packages.

The most common method of removing large amounts of waste heat in a laser diode package is by using microchannel-cooled packaging technologies. This method allows for cooling fluid to pass very near to the laser diode bar, with typical distances from the laser diode bar to the cooling fluid of approximately 200 μm. Most commercially-available microchannel coolers are made by diffusion bonding multiple layers of copper, each of which is typically 200-400 μm thick. The laser diode bar is often soldered directly to the copper cooler with indium solder. It is also common to solder the bar to a CTE-matched heatsink, and then solder the resulting subassembly to a copper MCC. This configuration allows hard solder to be more easily used.

While this approach provides excellent thermal performance, there are several drawbacks. The use of metallic microchannel coolers causes the electrical path to come in direct contact with the coolant. This requires the use of deionized water in order to ensure that no electricity flows through the coolant lines. Typically, water with a resistivity on the order of 0.5 MOhm*cm is used to cool microchannels in laser diode arrays.

The use of deionized water in microchannel-cooled laser diode arrays has led to a number of well-documented failure mechanisms, most notably the erosion and corrosion of the microchannel coolers [1]. Minimization of the corrosion failure mechanism requires a great deal of attention be paid to the entire cooling system [2]. In particular, it requires that the chiller be equipped with a means of controlling the pH and resistivity of the water. The use of certain common plumbing materials (i.e. brass) is also forbidden.

Erosion of copper microchannel coolers is also a significant problem. The thermal performance of a copper MCC can typically be improved by increasing the coolant flow rate. However, this increases the water speed through the cooler channels and speeds up the erosion process. Therefore two of the most critical factors in overall laser diode array lifetime – the thermal performance and robustness of the package – have opposing relationships to increased device flow rate.

While these hurdles have been overcome in many applications, they present substantial barriers in others. The additional cost and size associated with chillers manufactured for deionized water applications makes their use cost prohibitive in many applications that could make use of small, microchannel-cooled arrays. In these applications the cost of the chiller can significantly outweigh the cost of the arrays. In addition, the complexity and maintenance needs of the cooling system place additional burdens on the end user of the laser system. This can be a roadblock to the success of the end user.

DESIGN

NGCEO has developed a new laser diode array package which possesses many of the benefits of copper MCCs but eliminates the drawbacks [3,4]. The new package is based on Low Temperature Cofired Ceramic (LTCC) technology. Low-Temperature Co-fired Ceramic technology has been used successfully in the electronics industry for many years. In particular, this technology has been used extensively in the manufacture of radio frequency devices, including many RF circuits in cell phones. This technology has also been used in a variety of microfluidic applications.

A side-view schematic of one of these coolers is shown in Figure 1. Water flows through cooling channels in the LTCC and impinges directly on the back of the heat spreader. These devices use electrically isolating submounts (AlN, BeO, or diamond) as heat spreaders in order to eliminate the need for deionized water and decrease the thermal resistance of the package. Since the coolant is in direct contact with the submount, the low thermal conductivity of the LTCC material does not adversely affect the thermal performance of the device.

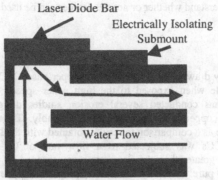

Figure 1. Side-view schematic of an LTCC MCC.

The materials used in these coolers offer several advantages over copper. First, all materials are significantly harder than copper which improves their erosion resistance. Second, the materials are closely CTE matched to GaAs. This allows for the use of hard solders, such as AuSn, in the assembly process. The use of AuSn solder has been repeatedly shown to improve device lifetimes when compared to devices built with soft solder (e.g. indium).

DEVICE FABRICATION

NGCEO has leveraged the conventional thick film processing and bonding techniques of the LTCC industry to create multi-layer devices with water channels similar in nature to copper MCCs. A typical copper MCC has five layers. The LTCC designs proposed by NGCEO have between three and nine layers. The preliminary samples used in this project were built by the Electrical Engineering department at the University of Idaho.

Since one of the primary goals of this project was to develop a MCC that could be easily fabricated by conventional LTCC manufacturers, every attempt was made to use industry-standard manufacturing methods. DuPont 951 green tape was nibbled using a CNC punching system to form the individual layers. Since these devices contain several large channels for fluid flow, a sequential lamination process was developed to minimize the occurrence of cavity collapse. This was the primary deviation from standard ceramic manufacturing practice used in the creation of these MCCs. In addition, a dicing saw was used to generate some of the more precise geometric requirements of the package. Further details can be found in previous works by the authors [5,6].

RESULTS

There are three main criteria that have to be evaluated when considering a new MCC. First, the thermal performance of the new package must be shown to be similar to (or better than) existing MCC designs. The predicted and measured thermal performance of these coolers has been reported in previous work [3,4]. Second, the long-term erosion and corrosion performance of the package has to be evaluated. Third, practical considerations (such as form and fit) must be considered in order to understand whether or not the devices can be used as direct replacements in existing systems.

Erosion Resistance

One of the primary drawbacks of the existing copper-based MCC technology is the fact that the coolers can erode when exposed to the high water speeds common to laser diode applications. NGCEO has conducted several erosion studies designed to understand the parameter space in which copper coolers can be operated reliably. The results of one such study are presented here to serve as a comparison to results obtained with the LTCC MCCs. A vertical stack of six copper MCCs was subjected to a flow rate of 0.2 GPM/cooler, which is approximately four times greater than the flow rate recommended by the manufacturer of the copper coolers. This was purely a test of the erosion properties of the coolers – no voltage or current was applied to the diode bars in the stack in order to eliminate the effects of galvanic corrosion.

At the conclusion of the erosion test, the top (mounting surface) layer of one of the coolers was removed and compared to an unused cooler. A picture of each cooler is shown in Figure 2. The water flow is such that it comes up through small holes near the front of the cooler (out of the page in Figure 2) and then away from the front of the cooler as indicated by the blue arrows. The zigzag pattern in the cooler is designed to promote turbulent flow and therefore improve the cooling properties of the MCC.

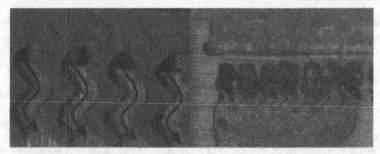

Figure 2. Examples of the internal structure of a copper MCC before (top) and after (bottom) a high-flow-rate test.

The effect of the high flow rate is clearly seen in the right hand side of Figure 2. Most of the structure that existed in the front of the cooler has been completely eroded by the water. The small holes in the front of the cooler that transport the water to the cooling layer have greatly expanded in size, and a significant portion of the zigzag pattern has been eroded away. This picture is representative of what has been observed in other tests conducted at high flow rates.

A similar test has been conducted with three ceramic coolers, one cooler with each of the thermal window materials (AlN, BeO, and CVD diamond). These coolers were packaged into single-bar MCC arrays and placed in a life test system. A flow rate of 0.25 GPM/cooler was set for the devices and the system was allowed to run without any pH or resistivity control. After approximately 550 hours, the sample with a BeO thermal window was removed from the system. The thermal window was removed and the internal channels of the cooler were examined under a microscope. The bottom of the thermal window was also examined to look for signs of erosion from the impinging water. The other two samples were tested for an a total run time of approximately 2500 hours, at which time the sample with the AlN thermal window was removed for a similar analysis. Photographs of the internal structure of these two samples, along with an unused cooler, are shown in Figure 3. (Note: The different color seen in the third image is a function of the microscope and camera settings. There was no change in the color of the device over the course of the test.)

Figure 3. Internal structure of example LTCC MCCs at three different times during a high-flow-rate test: before the test (top), 550 hours (middle), 2500 hours (bottom).

The performance differences between the copper and ceramic coolers in high-flow conditions are striking. The ceramic coolers exhibit little (if any) erosion after 2500 hours at 0.25 GPM/cooler, whereas the copper coolers had eroded to the point of failure after 1000 hours at 0.2 GPM/cooler. Of equal importance is the erosion resistance of the thermal window materials. The BeO examined at 550 hours and the AlN examined at 2500 hours each showed very limited signs of erosion. This test demonstrates the robust nature of the ceramic microchannel coolers.

The resistance to erosion of the LTCC MCCs opens the door to a wide range of operating conditions. In applications where high flow rates are available, they can be used to improve the thermal performance of the devices with little negative impact on overall array lifetime.

Array Configuration

NGCEO has developed several different MCC designs based on the LTCC technology. One MCC is a direct form and fit replacement for existing copper MCCs produced by NGCEO. These MCCs can be stacked vertically and horizontally and result in arrays with the same bar-to-bar pitch as is standard for arrays based on copper MCCs. As a result, existing copper-based MCC arrays can be replaced with LTCC-based MCC arrays with little impact to the overall laser system. An example six-bar LTCC MCC array is shown in Figure 4.

Figure 4. Six-bar array built from LTCC MCCs. This array is a form-and-fit replacement for a six-bar array built from copper MCCs.

One of the primary strengths of this approach is the high level of configurability that is possible with the LTCC MCCs. The manufacturing process for multilayer LTCC substrates is very mature. As a result there is little additional lead time associated with custom MCC designs. This enables NGCEO to design coolers which meet the exact specifications of each customer's application.

CONCLUSIONS

Northrop Grumman Cutting Edge Optronics has developed a new laser diode cooling technology using ceramic microchannel coolers. These coolers have many of the same strengths as the current state-of-the-art copper microchannel coolers, but they do not share the same weaknesses. Since the thermal and electrical paths are electrically isolated, standard filtered water can be used as a coolant (deionized water is not required). In addition, the robust nature of the ceramic materials used enables a much higher degree of erosion resistance. The design flexibility also enables the creation of direct replacements for many copper coolers and copper-based diode arrays currently on the market.

REFERENCES

1. Georg Truesch et. al., "Reliability of Water Cooled High Power Diode Laser Modules," proc. of SPIE 5711, 132-141 (2005).
2. John Haake and Brian Faircloth, "Requirements for Long Life Micro-Channel Coolers for Direct Diode Laser Systems," proc. of SPIE 5711, 121-131 (2005).
3. Ryan Feeler et. al., "Elimination of Deionized Cooling Water Requirement for Microchannel-Cooled Laser Diode Arrays," proc. of SPIE 6456, 645617 (2007).
4. Ryan Feeler et. al., "Next-Generation Microchannel Coolers," proc. of SPIE 6876 (2008).
5. Edward Stephens et. al., "Micro-fluidic Optoelectronic Packages based on LTCC," proc. of 2007 International Microelectronics and Packaging Conference, 429-436 (2007).
6. Fred Barlow et. al., "Fabrication of Precise Fluidic Structures in LTCC", proc. of CICMT (2008) (submitted).

Photodetection Devices

Mater. Res. Soc. Symp. Proc. Vol. 1076 © 2008 Materials Research Society   1076-K02-02

Single Carrier Initiated Low Excess Noise Mid-Wavelength Infrared Avalanche Photodiode using InAs-GaSb Strained Layer Superlattice

Koushik Banerjee[1], Shubhrangshu Mallick[1], Siddhartha Ghosh[1], Elena Plis[2], Jean Baptiste Rodriguez[2], Sanjay Krishna[2], and Christoph Grein[3]

[1]Lab for Photonics and Magnetics (ECE), University of Illinois at Chicago, Chicago, IL, 60607
[2]Center for High Technology Materials (ECE), University of New Mexico, Albuquerque, NM, 87106
[3]Microphysics Laboratory (Physics), University of Illinois at Chicago, Chicago, IL, 60607

ABSTRACT

Mid-wavelength infrared (MWIR) avalanche photodiodes (APDs) were fabricated using Indium Arsenide- Gallium Antimonide (InAs-GaSb) based strain layer superlattice (SLS) structures. They were engineered specifically to have either electron or hole dominated ionization. The gain characteristics and the excess noise factors were measured for both devices. The electron dominated p^+-n^--n APD with a cut-off wavelength of 4.13 μm at 77 K had a maximum multiplication gain of 1800 measured at -20 V while that of the hole dominated n^+-p^--p structure with a cut-off wavelength of 4.98 μm at 77 K was 21.1 at -5 V at 77 K. Excess noise factors were measured between 1-1.2 up to a gain of 800 and between 1-1.18 up to a gain of 7 for electron and hole dominated APDs respectively.

INTRODUCTION

Long range military and astronomical applications need to detect, recognize and track a variety of targets under a wide range of atmospheric conditions. This becomes particularly difficult due to significant absorption of the optical signal by the atmospheric gases (carbon-dioxide, carbon-monoxide and water vapor). In addition, the scattering by the air-borne dust particle makes it even more indiscernible. As a result, at the receiver, an amplification step becomes necessary in addition to the detection stage to get the detector signal above the preamp noise level. The APDs play a very significant role by combining both stages in a single device and thus simplifying the overall receiver complexity. However, avalanching being a random process, both in terms of photon-absorption and ionization, a high frequency noise known as the excess noise is incorporated at the output signal of an APD. This noise generally increases with the gain and often limits optimal performance of the device. Naturally, an APD having an excess noise factor minimally varying with gain is of prime importance for longer wavelength infrared detection. Mercury cadmium telluride (HgCdTe) has been the standard material of choice for such research in the last five to six years [1-8]. However, InAs-GaSb based superlattices, where novel band structures can be attained with the proper choice of the composition and thickness of individual layers, have also been a prospective candidate for the APD application. They have the additional advantages of relatively easier growth [9], comparable absorption coefficient [10] and longer Auger lifetime for p-doped sample [11-12], potentially resulting in an enhanced detectivity (D^*) [13], compared to HgCdTe detectors. Photodetectors, with performance comparable to HgCdTe photodetectors, have been designed, fabricated and characterized using InAs-GaSb SLS [14-16]. In our previous work [17-19], we have demonstrated the high gain and low noise characteristic of an electron dominated InAs-GaSb SLS APD. In this work, we present low excess noise factors over a higher gain regime for the electron dominated device. In addition, we designed and fabricated a new APD which has hole dominated avalanching and shows a noiseless characteristics.

THEORY AND DESIGN

The electronic band structure of the electron dominated superlattice (24 Å InAs/3 Å InSb/24 Å GaSb/3 Å InSb) having a bandgap of 300 meV at 77 K is shown in figure 1. The cut-off wavelength of 4.13 μm, as calculated from the band-gap, was verified from Fourier Transform Infrared (FTIR) spectroscopy [18]. When a reverse bias is applied to the device, owing to lower electron effective mass, electrons in CC1 band reach the higher energy states easily. When one of those electrons relaxes back to the bottom of CC1, the emitted energy is absorbed by another electron in HH1 and it is promoted to the bottom of CC1 leaving a hole in HH1. The flatness of the first heavy-hole (HH1) band, arising due to the special superlattice structure, aids the simultaneous conservation of energy and momentum in the process.

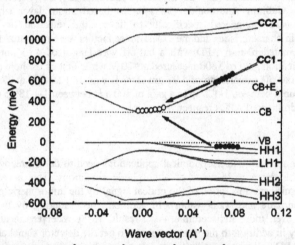

Figure 1: Band Structure of 24 Å InAs /3 Å InSb/24 Å GaSb/3 Å InSb at 77 K along with the most probable impact ionization.

Figure 2 shows the electronic band structure for hole-initiated avalanching (35 Å InAs/3 Å GaAs/20 Å Ga$_{60}$In$_{40}$Sb/13 Å AlSb/1 Å GaAs) having a bandgap of 248.84 meV at 77 K. The bandgap suggests a cut-off wavelength of 4.98 μm. In this structure the resonance occurs close to the zone center as the energy difference between LH1 and HH1 is almost equal to the bandgap. When a light hole in LH1 relaxes to the top of HH1, the emitted energy is absorbed by an electron on the top of HH1 and it is promoted to the bottom of CC1 leaving a hole in HH1. The electronic band structures were simulated using a 14 band k.p model. The positive wave vector represents the direction perpendicular to the superlattice growth.

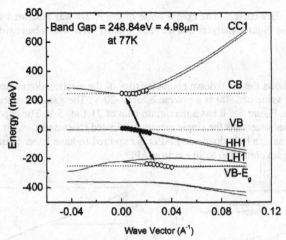

Figure 2. Band Structure of 35 Å InAs/3 Å GaAs/20 Å Ga$_{60}$In$_{40}$Sb/13 Å AlSb/1 Å GaAs at 77 K along with the most probable impact ionization.

EXPERIMENT

The structures were grown using molecular beam epitaxy (MBE) on a GaSb (001) substrate. The devices were characterized in the front illumination mode, so during the growth the p-contact was at the top and the n-contact at the bottom for electron dominated structure and vice-versa for the hole dominated one. The detail of the electron dominated structure is given in [18]. In the hole dominated avalanching structure, the superlattice absorber region is 1.44 μm thick with the superlattice structure repeated for 200 periods. It is doped p-type with Be having a doping density of 1×10^{15} cm^{-3}. Above and below the absorber region, there are n-type and p-type regions respectively which have the same constituent layers but repeated for 70 periods. In the p-contact, the GaSb layer was doped with Be and in the n-contact, the InAs layer was doped with GaTe.

Mesa structures were etched with 2:1:20 H$_2$O$_2$:H$_3$PO$_4$:DI water etching solution with continuous agitation. 4400 Å ZnS was deposited using electron-beam deposition as the passivation layer. The sample was treated with 20% warm aqueous solution of (NH$_4$)$_2$S for 10 minutes before passivation to reduce the interface state density at the passivant-SLS interface [19]. 200 Å Ti/200 Å Pt/4000 Å Au was used as the contact and pad metal.

The fabricated devices were characterized inside a Janis temperature and pressure controlled probe station. The electron APDs had a 400 μm square area while the hole APDs were circular with diameter 300 μm. Current-voltage characteristics were measured with the help of Agilent 4156B precision semiconductor parameter analyzer. The photocurrent and gain of the APDs were measured with a 632 nm He-Ne laser with an incident power of 300 μW. For the gain calculation, the difference between the photo-current and the dark current was normalized with the unity-gain dark current. The excess noise of the APDs was measured with the help of a Agilent 8973B Noise Figure Analyzer at a frequency of 10 MHz with a bandwidth of 4 MHz.

DISCUSSION

Figure 3a shows the breakdown characteristics of the electron dominated APD. The gain attains a maximum value of 1800 at a reverse bias of -20 V. The gain characteristics of the hole APD is presented in Figure 3b. It has a maximum gain of 21.1 at -5 V. The exponential nature of the gain in both cases ascertains the single carrier dominated characteristics. Since the holes have a higher effective mass, a hole dominated APD is expected to have lower ionization coefficient and hence a lower gain compared to an electron APD.

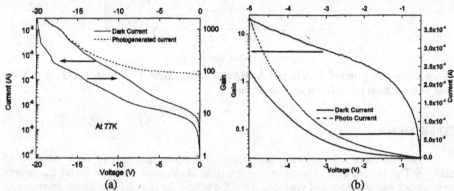

Figure 3. Photocurrent, dark-current and corresponding multiplication gain vs. applied reverse bias for a) electron APD and b) hole APD at 77 K.

Excess noise factors of 1-1.2 were measured up to a gain of 800 for the electron APDs as shown in Figure 4a. The same for hole APDs were measured between 1-1.18 up to a gain of 7 as shown in Figure 4b. Both structures exhibit low excess noise close to 1 which does not vary much with increasing gain. According to McIntyre theory [20], the excess noise is low when the avalanching process is dominated by a single carrier and the avalanching process is initiated by the dominating carrier. Since the front illuminated p^+-n^--n (n^+-p^--p) structure for electron (hole) APD ensures primarily electron (hole) initiated avalanching, the low excess noise factors are clear indication of electron (hole) dominated avalanching. A comparison of the experimental values with those from McIntyre formula (as illustrated in Figures 4a and b) leads to the same conclusion. However, McIntyre theory suggests that the minimum value of the excess noise can be 2 which is higher than the experimental values. This anomaly can be explained from dead space theory [21-22]. A more detailed explanation can be found in [18].

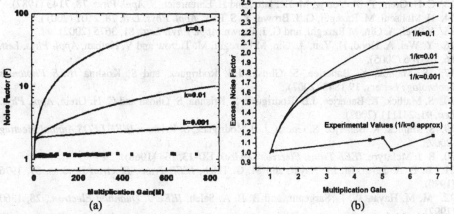

Figure 4. Excess noise factor vs. multiplication gain characteristics for a) electron initiated APD b) hole initiated APD along with theoretical McIntyre values (k is the ratio of hole to electron ionization coefficient).

CONCLUSIONS

In conclusion, avalanche photodiodes were fabricated on InAs-GaSb superlattices which were specifically lattice engineered to have either electron only or hole only avalanche mechanism. Both APDs showed very low excess noise with minimal change with multiplication gain. The electron APD had a maximum gain of 1800 at -20 V with excess noise factor 0.8-1.2 up to a gain of 800. The hole APD had a maximum gain of 21.1 at -5 V with excess noise factors 1-1.18 up to a gain of 7.

REFERENCES

1. M. A. Kinch, J. D. Beck, C-F Wan, F. Ma and J. Campbell J. Electronic Mat. 33 630 (2004).
2. J. D. Beck, C-F. Wan, M. Kinch, J. Robinson Proc. SPIE 4454, 188 (2001).
3. J. Beck, C. Wan, M. Kinch, J. Robinson, P. Mitra, R. Scritchfield, F. Ma, and J. Campbell J. Elec. Mat. 35 1166 (2006).
4. M. B. Reine, J. W. Marciniec, K.K. Wong, T. Parodos, J. D. Mullarkey, P. A. Lamarre, S. P. Tobin, K. Gustavsen, and G. Williams   Proc. SPIE 6294, 629403 (2006).
5. G. Destefanis, and P. Tribolet Proc. SPIE 6542, 723467 (2007).
6. G. Perrais,O. Gravrand, J. Baylet, G. Destefanis, and J. Rothman J. Elec. Mat. 36, 963 (2007).
7. S. Mallick, S. Ghosh, S. Velicu, J. Zhao, in press *IEEE Journal of Electronic Material*.
8. S. Mallick, S. Ghosh, S. Velicu, J. Zhao, *Proc. of SPIE*, 6660, 66600Y (2007)
9. G. C. Osbourn, *J. Vac. Sci. Technol.*, B2, 176 (1984).
10. D. L. Smith, and C. Mailhiot, *J. Appl. Phys.*, 62, 2545 (1987).
11. C. H. Grein, P. M. Young, and H. Ehrenreich, *Appl. Phys. Lett,.* 61, 2905 (1992).
12. E. R. Youngdale, J. R. Meyer, C. A. Hoffman, F. J. Bartoli, C. H. Grein, P.M. Young, H. Ehrenreich, R. H.Miles, and D. H. Chow, *Appl. Phys. Lett.,.*64,.3160 (1994).

13. C. H. Grein, P. M. Young, M. E. Flatte, and H. Ehrenreich, *J. Appl. Phys.*,**78**, 7143 (1987).
14. H. Mohseni, M. Razeghi, G. J. Brown, Y. S. Park, *Appl. Phys. Lett*, **78**, 2107 (2001).
15. Y. Wei, A. Gin, M Razeghi, and G. J. Brown, *Appl. Phys. Lett.*, **81**, 3675 (2002).
16. Y. Wei, A. Hood, H. Yau, A. Gin, M Razeghi, M. Tidrow and V. Nathan, *Appl. Phys. Lett*, **86**, 233106 (2005).
17. S. Mallick, K. Banerjee, S. Ghosh, J.B. Rodriguez, and S. Krishna *IEEE Photonics Technology Letters,* **19** 1843 (2007).
18. S. Mallick, K. Banerjee, J.B. Rodriguez, S. Krishna, S. Ghosh and C. H. Grein, *Appl. Phys. Lett*, **91**, 241111 (2007).
19. S. Mallick, K. Banerjee, S. Ghosh, J. B. Rodriguez, S. Krishna, *IEEE LEOS Annual Meeting -2007.*
20. R. J. McIntyre, *IEEE Trans. Electron Devices,* **ED-13**, 164 (1966)
21. B. E. A. Saleh, M. M. Hayat, and M. C. Teich, *IEEE Trans. Electron Devices*, **37**, 1976 (1990).
22. M. M. Hayat, W. L. Sargeant, and B. E. A. Saleh, *IEEE J. Quantum Electron.*, **28**, 1360 (1992).

Mater. Res. Soc. Symp. Proc. Vol. 1076 © 2008 Materials Research Society   1076-K03-02

Quantum Well and Quantum Dot Based Detector Arrays for Infrared Imaging

Sarath Gunapala, Sumith Bandara, Cory Hill, David Ting, John Liu, Jason Mumolo, Sam Keo, and Edward Blazejewski
Jet Propulsion Laboratory, California Institute of Technology, 4800 Oak Grove Drive, Pasadena, CA, 91109

ABSTRACT

Mid-wavelength infrared (MWIR) and long-wavelength infrared (LWIR) 320x256 pixel quantum well infrared photodetector (QWIP) dualband focal plane arrays (FPAs) have been demonstrated with excellent imagery. Currently, we are developing a 1024x1024 pixel simultaneous pixel co-registered dualband QWIP FPA. In addition, epitaxially grown self-assembled InAs/InGaAs/GaAs quantum dots (QDs) are exploited for the development of large-format FPAs. The Dot-in-a-Well (DWELL) structures were experimentally shown to absorb both $45°$ and normal incident light, therefore a reflection grating structure was used to enhance the quantum efficiency. The devices exhibit peak responsivity out to 8.1 microns, with peak detectivity reaching $\sim 1 \times 10^{10}$ Jones at 77 K. The devices were fabricated into the first LWIR 640x512 pixel QDIP FPA, which has produced excellent infrared imagery with the NETD of 40 mK at 60K operating temperature.

INTRODUCTION

There are many applications that require mid-wavelength infrared (MWIR) and long-wavelength infrared (LWIR) dualband focal plane arrays (FPAs). For example, a dualband FPA camera would provide the absolute temperature of a target with unknown emissivity, which is extremely important to the process of identifying a temperature difference between missile targets, warheads, and decoys. Dualband infrared FPAs can also play many important roles in Earth and planetary remote sensing, astronomy, etc. Furthermore, monolithically integrated pixel collocated simultaneously readable dualband FPAs eliminate the beam splitters, filters, moving filter wheels, and rigorous optical alignment requirements imposed on dualband systems based on two separate single-band FPAs or a broadband FPA system with filters. Dualband FPAs also reduce the mass, volume, and power requirements of dualband systems. Due to the inherent properties such as narrow-band response, wavelength tailorability, and stability (i.e., low 1/f noise) associated with GaAs based QWIPs [1-6], it is an ideal candidate for large format dualband infrared FPAs.

320X256 PIXEL DUALBAND FOCAL PLANE ARRAY

As shown in figure 1, our dualband FPA is based on two different types of (i.e., MWIR and LWIR) QWIP devices separated by a 0.5 micron thick, heavily doped, n-type GaAs layer. The device structures of the MWIR and LWIR devices are very similar to the MWIR and LWIR The MWIR device described here, each period of the multi-quantum-well (MQW) structure consists of coupled quantum wells of 40 Å containing 10 Å GaAs, 20 Å $In_{0.3}Ga_{0.7}As$, and 10 Å GaAs layers (doped n = 1×10^{18} cm^{-3}) and a 40 Å undoped barrier of $Al_{0.3}Ga_{0.7}As$ between coupled quantum wells, and a 400 Å thick undoped barrier of $Al_{0.3}Ga_{0.7}As$. Each period of the

LWIR MQW structure consists of quantum wells of 40 A and a 600 A barrier of $Al_{0.27}Ga_{0.73}As$. Stacking many identical periods (typically 50) together increases photon absorption. Both device structures and heavily doped contact layers were grown in-situ during a single growth run using molecular beam epitaxy. It is worth noting that the photosensitive MQW region of each QWIP device is transparent at other wavelengths, which is an important advantage over conventional interband detectors. This spectral transparency makes QWIPs ideally suited for dualband FPAs with negligible spectral cross-talk. As shown in figure 1, the carriers emitted from each MQW region are collected separately using three contacts. The middle contact layer is used as the detector common. The electrical connections to the detector common, and the LWIR pixel connection, are brought to the top of each pixel using via connections.

Light coupling to a pixel collocated dualband QWIP device is a challenge since each device has only a single top surface area. We have developed two different optical coupling

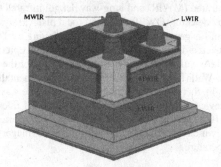

Figure 1. 3-D view of dualband QWIP device structure showing via connects for independent access of MWIR and LWIR devices.

techniques. The first technique uses a dual period Lamar grating structure. The second technique uses the multiple diffraction orders. In this light coupling technique, we have used a 2-D grating with single pitch. The first diffraction orders (1,0), (0,1), (-1,0), and (0,-1) couple infrared radiation into LWIR pixels, and the second diffraction orders (1,1), (-1,1), (1,-1), and (-1,-1) couple infrared radiation into MWIR pixels. The spectral responsivity of dualband QWIP is shown in figure 2. 2-D periodic grating structures were designed to couple the 4-5 and 8-9 µm radiation into the detector pixels. The top 0.7 µm thick GaAs cap layer was used to fabricate the light coupling 2-D periodic grating. The 2-D grating reflectors on top of the detectors were then covered with Au/Ge and Au for Ohmic contact and reflection.

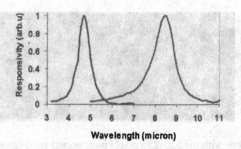

Wavelength (micron)

Figure 2. Responsivity of the dualband QWIP device as a function of wavelength.

After the 2-D grating array was defined by photolithography and dry etching, the MWIR detector pixels of the 320x256 pixel FPAs, and the via hole to access the detector common, were fabricated by dry etching through the photosensitive GaAs/In$_y$Ga$_{1-y}$As/Al$_x$Ga$_{1-x}$As MQW layers into the 0.5 µm thick doped GaAs intermediate contact layer. Then LWIR pixels and via holes to access the LWIR pixels of FPAs were fabricated. A thick insulation layer was deposited and contact windows were opened at the bottom of each via hole and on the top surface. Ohmic contact metal was evaporated and unwanted metal was removed using a metal lift-off process. The pitch of the FPA is 40 µm and the actual MWIR and LWIR pixel sizes are 38x38 µm^2. Forty eight FPAs were processed on a four-inch GaAs wafer. Indium bumps were then evaporated on top of the detectors for silicon read out integrated circuit (ROIC) hybridization. Several dualband FPAs were chosen and hybridized (via an indium bump-bonding process) to a 320x256 pixel CMOS read out integrated circuit. A 320x256 pixel co-registered simultaneously readable dualband QWIP FPA hybrid was mounted onto a 5 W integral Sterling closed-cycle cooler assembly to demonstrate a portable MWIR:LWIR dualband QWIP camera. The digital acquisition resolution of the camera is 14-bits, which determines the instantaneous dynamic range of the camera (i.e., 16,384). However, the dynamic range of QWIP is 85 Decibels. Video images were taken at a frame rate of 30 Hz at temperatures as high as T = 68 K, using two ROIC capacitors having charge capacities of 21x10E6 and 87x10E6 electrons for the MWIR and LWIR bands respectively. Figure 3 shows an image taken with the 320x256 pixel co-registered simultaneously readable MWIR:LWIR dualband QWIP camera. As expected (due to background limited performance [BLIP]), the estimated and experimentally obtained the NEΔT values of the LWIR detectors do not change significantly at temperatures below 65 K. The estimated NEΔT of MWIR and LWIR detectors at 65 K are 22 and 24 mK, respectively. These estimated NEΔT values based on the test detector data agree reasonably well with the experimentally obtained values. The experimentally measured NEΔT values are shown in the figure 4 (a) and (b). The experimentally measured NEΔT values are slightly higher than the estimated NEΔT value based on the results of single element test detector data. This degradation in signal-to-noise ratio is attributed to the inefficient light coupling of the dual feature lamalar grating coupler, an unoptimized ROIC, and the significant amount of 1/f noise in the FPA characterization equipment.

Figure 3. An image taken with the 320x256 pixel co-registered simultaneously readable MWIR:LWIR dualband QWIP camera. The person in this image is holding a cigarette lighter. The cigarette lighter produced lots of hot CO_2 gas. So, the flame in the MWIR image looks broader due to re- emission of the infrared signal in 4.1-4.3 microns band by the heated CO_2 gas produced by the cigarette lighter. Whereas, the heated CO_2 gas doesn't have any emission line in the LWIR (8-9 microns) band. Thus, the LWIR image shows only thermal signatures of the flame. The hot cigarette lighter flame produced a lot of MWIR signal, so it reflects off from the lens and face.

As we have mentioned earlier, QWIP is an ideal detector for the fabrication of pixel co-registered simultaneously readable dualband infrared focal plane arrays, because, QWIP absorbs infrared radiation only in a narrow spectral band which is designed to do so, and the transparent outside of that absorption (i.e., detection) band.. Thus it provides zero spectral cross-talk when two spectral bands are a few microns apart. The initial GaAs substrate of these dualband FPAs are completely removed leaving only a 50 nm thick GaAs membrane. Thus, these dualband QWIP FPAs are not vulnerable to FPA delamination and indium bump breakage during thermal recycling process, and has no pixel-to-pixel optical cross-talk. Inspired from this success, now we are developing a megapixel (1024x1024 pixel) dualband QWIP FPA sensitive in MWIR and LWIR spectral bands.

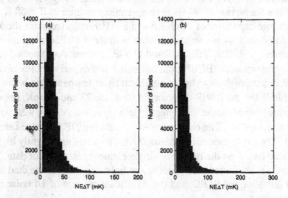

Figure 4. NEΔT histogram of the 320x256 format simultaneously readable pixel co-registered dualband QWIP focal plane array. Each spectral band of the FPA consisted of 320x256 co-registered pixels. The experimentally measured NEΔT of MWIR and LWIR detectors at 65 K are 28 and 38 mK, respectively.

DEVELOPMENT OF 1024X1024 PIXEL MWIR AND LWIR DUALBAND QWIP FPA

Most infrared FPAs consist of non-silicon detector arrays and silicon ROICs. Silicon ROICs are usually fabricated on large area (i.e., 8 – 12-inch dia.) wafers. In the process of large format IR FPA development, it is necessary to select an IR detector technology based on large area wafers. QWIP FPA technology is entirely based on the highly stable III-V material system that can be easily processed with the more mature fabrication technologies.

The state-of-the-art array fabrication processes are based on either mask aligners or reticle-based steppers. A typical reticle field is 22x22 mm^2. The pixel pitch of the largest 1Kx1K array fits in to a reticle field that is 18 µm. The pixel pitch of the 1Kx1K pixel dualband QWIP FPA is 30 µm. Thus, a 30 µm pixel pitch 1Kx1K array cannot be fabricated using conventional reticle stepping or mask aligning methods. As shown in figure 5, a large detector array can be fabricated using "Stitching". Stitching is a new photolithographic technique that can be used to fabricate detector arrays larger than the reticle field of photolithographic steppers. We have used the stitching technique to fabricate 1Kx1K arrays that can be easily extended into the fabrication of 2Kx2K and 4Kx4K detector arrays. In this case, the detector array layout is divided into smaller portion "tiles", which together fit in the reticle field. Array characteristics or repeated sections of the detector array are exploited to minimize the required reticle area by using multiple exposures of smaller blocks to create a large array. Each detector array is then photocomposed on the wafer by multiple exposures of detector array sections at appropriate locations on the wafer. Single sections of the detector array are exposed at one time, as the optical system allows shuttering, or selectively exposing only a desired section of the reticle. Figure 5 depicts photocomposition of the detector array on a wafer by stitching. It should be noted that stitching creates a truly seamless detector array, as opposed to an assembly of closely butted pieces.

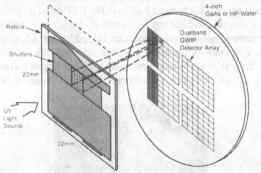

Figure 5. This figure shows the photocomposition of a detector array die using array stitching based on a photolithographic stepper.

After the 2-D grating array was defined by stepper based photolithography and dry etching, the MWIR detector pixels of the 1024x1024 pixel detector arrays, and the via hole to access the detector common, were fabricated by dry etching through the photosensitive GaAs/In$_y$Ga$_{1-y}$As/Al$_x$Ga$_{1-x}$As MQW layers into the 0.5 µm thick doped GaAs intermediate contact layer. Then LWIR pixels and via holes to access the LWIR pixels of FPAs were fabricated. A thick insulation layer was deposited and contact windows were opened at the bottom of each via hole and on the top surface. Ohmic contact metal was evaporated and unwanted metal was removed using a metal lift-off process. The pitch of the FPA is 40 µm and the actual MWIR and LWIR pixel sizes are 28x28 µm^2. Five detector arrays were processed on a four-inch GaAs wafer. Indium bumps were then evaporated on top of the detectors for hybridization with ROICs. Several dualband detector arrays were chosen and hybridized (via an indium bump-bonding process) to grade B (i.e., some dead columns) 1024x1024 pixel silicon ROICs.

Array thinning or the substrate removal process is critical to the success and durability of large format cryogenic FPAs. Thus, after the detector array and ROIC hybridization process via indium bumps, the gaps between FPA detectors and the ROIC are backfilled with epoxy. This epoxy backfilling provides the necessary mechanical strength to the detector array and ROIC hybrid prior to the thinning process. During the first step of the thinning process, an approximately 630 µm thick GaAs layer was removed using diamond point turning. Then Bromine-Methanol chemical polishing was used to remove another approximately 100 µm thick GaAs layer. This step is very important because it removes all scratch marks left on the substrate due to abrasive polishing. Otherwise these scratch marks will be enhanced and propagated in to the final step via preferential etching. Then, wet chemical etchant was used to reduce the substrate thickness to several microns and then SF_6:BCl_3 selective dry etchant was used as the final etch. This final etching completely removed the remaining GaAs substrate. At this point the remaining GaAs/AlGaAs material contains only the QWIP pixels and a very thin membrane (~500Å). The thermal mass of this membrane is insignificant compared to the rest of the hybrid. This allows it to adapt to the thermal expansion and contraction coefficients of the silicon ROIC and completely eliminates the thermal mis-match problem between the silicon based readout and the GaAs based detector array. This basically allows QWIP FPAs to go through an unlimited number of temperature recycles without any indium bump breakage and array delamination. Furthermore, this substrate removal process provides two additional advantages for QWIP FPAs: those are the complete elimination of pixel-to-pixel optical cross-talk and a significant (a factor of two with 2-D periodic gratings) enhancement in optical coupling of infrared radiation into QWIP pixels. Figure 6 shows a megapixel dualband QWIP FPA mounted on a 124 pin LCC.

Figure 6. Picture of a 1024x1024 pixel silicon ROIC mounted on a 100-pin lead less chip carrier (LLC).

Primary goals of these test hybrids were to test their mechanical stability, sub-strate removal, and pixel connectivity. A test (i.e., used grade B RIC) MWIR:LWIR pixel co-registered simultaneously readable dualband QWIP FPA has been mounted on to the cold finger of a pour fill dewar, cooled by liquid nitrogen and the two bands (i.e., MWIR and LWIR) were independently biased. Some imagery was performed at temperature 68 K. An image is shown in figure 7. The left panel shows a MWIR image and right panel shows a LWIR image of a person holding a hair dryer. The poor image quality of the MWIR image is due to poor pixel connectivity (~ 60%), which attributes to the breakage of metal connectors in via holes. Figure 8 shows some SEM pictures of the via connects, which connects the bottom of MWIR pixels to the bottom detector common under the LWIR pixels. The left panel shows the connectivity of this via metal connectors, however, the right panel shows some broken metal connectors. This confirms the poor pixel operability of MWIR is due to via metal breakage. These metal layers were deposited via e-beam metal evaporation. We think this metal breakage issue can be solved with sputtering based metal deposition due to its conformal coverage. During the next wafer run we should be able to improve the pixel operability and performance by using sputter deposition of via connect metal and grade A ROICs.

Figure 7. An image taken with the 1024x1024 pixel co-registered simultaneously readable MWIR:LWIR dualband QWIP camera. The left panel shows a MWIR image and the right panel shows a LWIR image of a person holding a hair dryer. Poor image quality of the MWIR image is due to poor pixel connectivity (~ 60%), which attributes to the breakage of metal connectors in via holes.

Figure 8. Scanning electron micrographs (SEMs) of via holes and metal bridges which connect the MWIR detector commons and LWIR detector common at the bottom of both pixels. The left panel shows good connectivity, whereas right panel shows disconnected metal bridges.

THE DOT-IN-THE-WELL INFRARED PHOTODETECTOR

The main benefit in using the QD approach stems from 3D quantum confinement, which (1) enables normal incidence absorption by modifying the optical transition selection rule, and, (2) increases the photo-excited carrier lifetime by reducing optical phonon scattering via the "phonon bottleneck" mechanism [7-10]. However, QDs also have some drawbacks that need to be addressed. In a typical detector structure, QD densities are low (compared to the number of dopants in the active regions of QWIPs); so while individual QDs are efficient absorbers, typical QD densities are not high enough to achieve high quantum efficiency. Thus, while QD based infrared detectors have clearly demonstrated normal incidence absorption, and, in some instances, higher operating temperature as well, they are still lacking in quantum efficiency and responsivity.

The first-generation QDIPs are high-gain, low-quantum-efficiency devices. Improving quantum efficiency is a key to achieving a competitive QD-based FPA technology. This can be accomplished by increasing the QD density, or by enhancing the infrared absorption in the QD-containing material. There are various versions of QDIPs, based on different materials and designs. After considering all types of competing QD-based approaches, we feel that one of the most promising options for LWIR FPAs is the use of the Dot-in-a-well (DWELL) QDIP. Our specific implementation uses InAs and InGaAs QDs embedded in GaAs/AlGaAs multi-quantum well structures, as illustrated in figure 9. This material system can sustain a large number of QD stacks without suffering material degradation, thereby allowing high dot density. The host GaAs/AlGaAs multi-quantum well structures are highly compatible with the mature FPA fabrication process that we use routinely to make QWIP FPAs; in that sense, this system may be viewed as simply a QWIP with embedded InAs QDs. Similar to other intersubband detectors, DWELLs operate by the photoexcitation of electrons between energy levels in the potential well created by the nano-scale QD in a well structure. The right panel of figure 9 shows that, under an applied bias voltage, these photo-excited carriers can escape from the potential wells and be collected as photocurrent. The wavelengths of the spectral peaks (λ_p) are determined by the energy difference between quantized states in the DWELL. The hybrid quantum-dot/quantum-well, or dot-in-a-well (DWELL), device offers two advantages: (1) challenges in the wavelength tuning through dot-size control can be compensated in part by engineering the quantum well sizes, which can be controlled precisely; (2) quantum wells can trap electrons and aid in carrier capture by QDs, thereby facilitating ground state refilling.

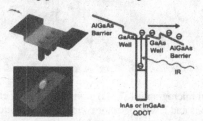

Figure 9. Illustration of the Dot-in-a-Well (DWELL) device. The top left panel shows the potential profile, with three pyramid shaped dots embedded in the quantum well. The bottom left panel displays a calculated DWELL ground state wave function, represented by a white translucent equal-probability isosurface, localized by a pyramidal quantum dot. The right panel illustrates the operation of a DWELL infrared detector.

MATERIAL GROWTH AND DEVICE FABRICATION

All DWELL-QDIP wafers were grown on semi-insulating 4" GaAs substrates using a Veeco Gen-III MBE Reactor. After evaluating the material quality, selected wafers were processed into test detector mesas. After the 200 µm x 200 µm square mesas are defined by lithography, the DWELL-QDIP test detectors were fabricated by standard wet and dry chemical etching through the stack of photosensitive layers into the doped GaAs bottom contact layer. The top contact of the detectors were covered with Au/Ge and Au for an Ohmic contact which also serves as a reflector for light incident through the bottom contact, allowing two passes through the active layers. Initial QDIP characterization of discrete devices included measurements of the room-temperature absorption spectra, side and normal incident responsivity spectra, dark current, and noise. These detectors were tested in a cryogenically cooled test bed using a calibrated blackbody source to evaluate responsivity of the detector over the relevant range of operating temperatures and bias voltages.

ELECTRICAL AND OPTICAL CHARACTERIZATION

During this development we have measured the normal and 45-degree incidence responsivity of DWELL QDIP samples. A typical set of results is shown in figure 10. As hoped, the normal incidence responsivity (relative to the 45-degree responsivity) is much stronger (almost 1 order of magnitude) than that found in a typical QWIP. At the same time, we also find that the 45-degree incidence responsivity is 4 to 5 times stronger than the normal incidence responsivity. This implies that at these wavelengths, the DWELL QWIPs not only have good absorption for normal incidence (x,y-polarized; with z being the normal incidence direction) light, it also absorbs inclined (or side) incidence (z-polarized) light even more strongly. This knowledge has prompted an important design modification, as we now wish to take advantage of the DWELL QDIP's ability to absorb both normal and inclined incidence light in order to maximize quantum efficiency. As in QWIPs, normal incidence light can be coupled to the z-polarization light absorption mechanism in DWELL QDIPs by using a reflection grating. Due to our extensive experience in designing and fabricating FPA-compatible integrated optical structures for QWIPs, we were in an excellent position to implement the grating reflector enhanced QDIP. Figure 11 shows our preliminary experimental results on grating enhancement. Normal incidence responsivity was measured for a DWELL QDIP sample, fabricated both with and without a reflection grating. The device with the reflection grating shows approximately three times larger normal incident responsivity than the one without, clearly indicating the promise of this approach.

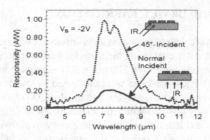

Figure 10. DWELL QDIP spectral responsivity measured for (a) normal incidence, (b) 45° incidence.

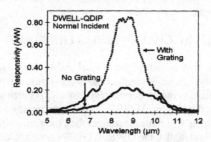

Figure 11. Normal incidence spectral responsivity of a DWELL-QDIP with and without reflection gratings.

Figure 12 shows the dark current as a function of temperature at various operating bias conditions. Figure 13 shows the D* of this detector as a function of device temperature at operating bias $V_B = -1V$. This figure shows that the DWELL detector reached background limited $D* \sim 1 \times 10^{11}$ Jones around T = 50 K temperature.

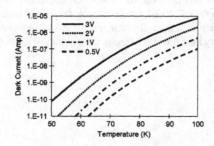

Figure 12. Dark current vs. operating temperature curves of a DWELL QDIP detector at various device bias conditions. The device area is 25x25 μm^2.

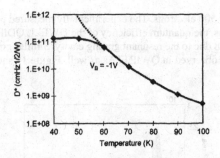

Figure 13. Dark current limited and background (300K, f/2 optics) limited specific detectivity of DWELL-QDIP as a function of operating temperature. Detectivities are calculated using experimentally measured spectral responsivity, dark current, and photoconductive gain of the detector.

QDIP FOCAL PLANE ARRAY RESULTS

After the 640x512 pixel QDIP detector arrays were hybridized to a 640x512 pixel ROIC, a simple electronic functionality test was performed to evaluate the FPAs. Selected detector hybrids were mounted and wire-bonded to a leadless chip carrier (LCC). A specially designed dewar was used to characterize the FPA functionality using a general-purpose electronic system from SE-IR Incorporated. The SE-IR system was programmed to generate complex timing patterns, and was reconfigured to handle development grade FPAs. The best performance was determined by optimizing operating-parameters for each FPA. The FPAs were characterized for pixel responsivity, quantum efficiency, noise, D^*, $NE\Delta T$, pixel operability, before and after correction uniformity, and pixel operability. The spectral responsivity of the FPA was determined using a separate single mesa processed with the FPA. $NE\Delta T$ as a function of bias and integration time at a fixed operating temperature was used as a metric for parameter optimization. Charge injection efficiency can be obtained from $E_{inj} = g_m R_D / (1 + g_m R_D)$, where gm is the trans- conductance of the MOSFET and is given by $g_m = eI_{Det}/kT$. The differential resistance R_{Det} of the 23x23 μm^2 pixels at -350 mV bias is 5.5×10^{10} Ohms at T = 60 K and detector capacitance C_{Det} is 1.4×10^{-14} F. The detector total current is I_{Det} = 17 pA under the same operating conditions. However, the dynamic resistance of the detector and ROIC is given by $R_D.C_{ROIC} = t_{int}$, where C_{ROIC} is the capacitance of the ROIC integration capacitor and t_{int} is the integration time. We have integrated the signal for 20 ms. The input capacitance of ISC 9803 ROIC is 350 fF, which yields R_D of 5.7×10^{10} Ohms. According to the equation above, the charge injection efficiency E_{inj}= 99.65% at a frame rate of 50 Hz. The FPA was back-illuminated through the flat thinned substrate membrane (thickness $\approx$ 1000 Å). This initial array gave very good images with >99% of the pixels working, demonstrating the high yield of GaAs technology. The operability was defined as the percentage of pixels having NEΔT within 2σ at 300 K background with f/2 optics and in this case operability happens to be equal to the pixel yield. Figure 14 shows the experimentally measured NEΔT histogram of the FPA at an operating temperature of T = 60 K, bias V_B = -350 mV at 300 K background with f/2 optics. The mean NEΔT value is 40 mK. This agrees reasonably well with our estimated value of 25 mK based on test structure data. The read noise of the multiplexer is 500 electrons. The experimentally measured peak quantum efficiency of the FPA was 5.0%. Thus, the quantum efficiency of the DWELL QDIP discussed in this paper enhanced by a factor of 1.8 due to the resonant grating cavity. Similar grating cavity enhancement effects were observed in QWIP FPAs as well. Figure 15 shows an image taken from this QDIP FPA.

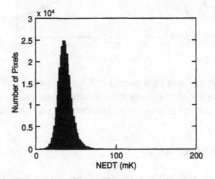

Figure 14. Noise equivalent temperature difference (NEΔT) histogram of the 311,040 pixels of the 640 x 512 pixel QDIP FPA showing a high uniformity of the FPA. The non-uniformity (= standard deviation/mean) of this unoptimized FPA is only 0.2%.

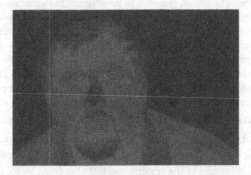

Figure 15. An image taken with the first 640x512 pixels QDIP LWIR focal plane array camera.

ACKNOWLEDGEMENTS

Authors are grateful to R. Cox, P. Dimotakis, M. Herman, E. Kolawa, A. Larson, R. Liang, T. Luchik, R. Odle, and R. Stirbl for encouragement and support during the development and optimization of QWIP FPAs at the Jet Propulsion Laboratory for various applications. The research described in this paper was performed by the Jet Propulsion Laboratory, California Institute of Technology, and was sponsored by the Missile Defense Agency and the Air Force Research Laboratory.

REFERENCES

1. H. C. Liu, F. Capasso (Eds.), Intersubband Transitions in Quantum Wells: Physics and Device Applications I and II, Academic Press, San Diego, 2000.
2. Sarath D. Gunapala, John K. Liu, Jin S. Park, Mani Sundaram, Craig A. Shott, Ted Hoelter, True-Lon Lin, S. T. Massie, Paul D. Maker, Richard E. Muller, and Gabby Sarusi "9 µm Cutoff 256x256 AlxGa1-xAs/AlxGa1–xAs Quantum Well Infrared Photodetector Hand-Held Camera", IEEE Trans. Electron Devices, 44, pp. 51-57, 1997.
3. Sarath D. Gunapala, Sumith V. Bandara, John K. Liu, Winn Hong, Mani Sundaram, Paul D. Maker, Richard E. Muller, Craig A. Shott, and Ronald Carralejo, "Long-Wavelength 640x486 GaAs/AlxGa1–xAs Quantum Well Infrared Photodetector Snap-shot Camera", IEEE Trans. Electron Devices, 45, 1890 (1998).
4. W. Cabanski, R. Breiter, R. Koch, K. H. Mauk, W. Rode, J. Ziegler, H. Schneider, M. Walther, and R. Oelmaier, "3rd gen Focal Plane Array IR Detection Modules at AIM", SPIE Vol. 4369, pp. 547-558 (2001).
5. S. D. Gunapala and S. V. Bandara, "Quantum Well Infrared Photodetector (QWIP) Focal Plane Arrays," Semiconductors and Semimetals, Vol. 62, 197-282, Academic Press, 1999.
6. S. D. Gunapala, S. V. Bandara, A. Singh, J. K. Liu, S. B. Rafol, E. M. Luong, J. M. Mumolo, N. Q. Tran, J. D. Vincent, C. A. Shott, J. Long, and P. D. LeVan, "640x486 Long-wavelength Two-color GaAs/AlGaAs Quantum Well Infrared Photodetector (QWIP) Focal Plane Array Camera" IEEE Trans. Electron Devices, Vol. 47, pp. 963-971, 2000.
7. V. Ryzhii, "The theory of quantum-dot infrared photodetector", Semicond. Sci. Technol., Vol. 11, pp. 759-765 (1996).
8. J. Phillips, K. Kamath, and P. Bhattacharya, "Far-infrared photoconductivity in delf-organized InAs quantum dots", Appl. Phys. Lett., Vol. 72, pp. 2020-2022 (1998).
9. S. D. Gunapala, S. V. Bandara, C. J. Hill, D. Z. Ting, J. K. Liu, S. B. Rafol, E. R. Blazejewski, J. M. Mumolo, S. A. Keo, S. Krishna, Y.-C. Chang, and C. A. Shott "640x512 pixels long-wavelength infrared (LWIR) quantum dot infrared photodetector (QDIP) imaging focal plane array", IEEE Journal of Quantum Electronics, Vol. 43, pp. 230 – 237 (2007).
10. S. Krishna, S. D. Gunapala, S. V. Bandara, C. Hill, and D. Z. Ting, "Quantum Dot Based Infrared Focal Plane Arrays", IEEE Special Issue on Optoelectronic Devices Based on Quantum Dots, 95, pp. 1838-1852 (2007).

Mater. Res. Soc. Symp. Proc. Vol. 1076 © 2008 Materials Research Society   1076-K03-04

Infrared Detector Activities at NASA Langley Research Center

M. Nurul Abedin[1], Tamer F. Refaat[2], Oleg V. Sulima[3], and Farzin Amzajerdian[4]
[1]RSFSB, NASA LaRC, 5 N. Dryden St., Hampton, VA, 23681
[2]Old Dominion University, Norfolk, VA, 23529
[3]University of Delaware, Newark, DE, 19716
[4]NASA LaRC, Hampton, VA, 23681

ABSTRACT

Infrared detector development and characterization at NASA Langley Research Center will be reviewed. These detectors were intended for ground, airborne, and space borne remote sensing applications. Discussion will be focused on recently developed single-element infrared detector and future development of near-infrared focal plane arrays (FPA). The FPA will be applied to next generation space-based instruments. These activities are based on phototransistor and avalanche photodiode technologies, which offer high internal gain and relatively low noise-equivalent-power. These novel devices will improve the sensitivity of active remote sensing instruments while eliminating the need for a high power laser transmitter.

INTRODUCTION

NASA's Earth Science Technology Office and Science Mission Directorate show great interest in broadband detectors for numerous critical applications such as temperature sensing, process control, and atmospheric monitoring of trace gases (CO_2, H_2O, CO, and CH_4). When selecting a detector for such applications, key parameters, must be considered in order to satisfy the requirements of the Earth and planetary remote sensing systems. These parameters include spectral response, quantum efficiency, noise-equivalent-power (NEP), and gain. In general, large format focal plane arrays (FPA) in the 1.0 to 2.5 µm spectral range are the detectors of choice for the next-generation Earth and Space Science remote sensing, Mars Orbiter, and planetary instruments. An Antimonide (Sb)-based heterojunction phototransistor (HPT), as a suitable candidate, stimulates a strong interest for broadband remote sensing applications [1-4]. Besides p-i-n photodiodes and avalanche photodiodes (APD), HPTs have also attracted a great attention to satisfy many of the detector requirements for these applications without excess noise and high bias voltage problems, while maintaining superior NEP. HPT has an internal gain mechanism that allows increasing the output signal and signal-to-noise ratio (SNR). However, further reduction of gain related noise is desirable, and research on different HPT structures indicate some important design considerations that can minimize device noise and further increase the SNR.

On the other hand, APDs are integrated solid-state semiconductor devices that manufactured using different materials. The common near-IR InGaAs/InP APDs are operational in the spectral range of 0.9 to 1.6 µm [5-7]. Recently, a group from DRS Technologies demonstrated responsivity of ~ 15 A/W using HgCdTe electron (e-) APD at short wave infrared (SWIR) wavelength (at -7 V and 20°C) [7]. However, HgCdTe e-APDs have demonstrated very high responsivity in the 3.0 to 5.0 μm range and their SWIR responsivity was low [7]. Even though HgCdTe detectors are efficient in the infrared range, significant technology limitations curb their application in that range. Therefore, researchers have been looking to other III-V systems that provide much higher yield by manufacturing and higher operability. In this context, it is necessary to explore alternative material systems, which can absorb radiation in the 1.0 to 2.5 μm range and are compatible with general requirements of a low-noise detector. InGaAsSb/AlGaAsSb heterostructures are very suitable for this application, and single-element HPTs based on these heterostructures demonstrated record responsivitites [3, 4, 8-14].

Quaternary InGaAsSb/AlGaAsSb HPTs in the 1.0 to 2.5 μm wavelength range have been developed and fabricated at AstroPower, Inc. in collaboration with NASA Langley Research Center (LaRC). These devices have been characterized at NASA LaRC and encouraging results have been obtained, including high responsivity, high detectivity, and relatively low NEP [3,4, 8-14]. Development of large-format Sb-based phototransistor FPA will enhance the capability of space-based active and passive remote sensing imaging systems and enable major advances in Astronomical, Earth, and planetary instruments. This new detector will enhance the capability of active remote sensing of CO_2 and water vapor at 2.05 and 1.9 μm, respectively.

There is no commercially available broadband detector with sufficient sensitivity in the 1.0 to 2.5 μm spectral range, whereas this spectral range is suitable for sounding Earth and planetary atmospheres. Therefore, we propose to develop the capability to reliably fabricate detector arrays that respond to broad near-IR wavelength regions.

DEMONSTRATION OF SINGLE ELEMENT INFRARED DETECTOR

Recently, NASA LaRC characterized single-element InGaAsSb/AlGaAsSb HPTs with a 200 μm diameter. Figure 1 shows the responsivity variation with wavelength in the 0.6 to 2.4 μm range at different collector-emitter voltages and temperatures of such a device [11]. The responsivity is strongly dependent on wavelength, temperature, and bias voltage. A maximum responsivity of 10000 A/W was achieved at 1.9 μm, -193°C and 5V. That was the highest responsivity ever reported in the 0.6 to 2.4 μm wavelength range for the Sb-based photodetectors.

Figure 2 compares the NEP variation with bias voltage of a HPT and of commercial InGaAs p-i-n photodiodes (Hamamatsu; G5852 and G5853) at different wavelengths [10]. The photodiodes have different cutoff wavelengths: 2.6 μm for G5853 and 2.3 μm for G5852. It was determined that at 2.0 μm, NEP of G5852 (top solid lines at -20 °C) is about five times lower compared to G5853. Comparison with the best of the tested p-i-n photodiodes shows that at 4.0 V the HPT achieves near an order of magnitude lower NEP despite the 40 degrees higher operating temperature (20°C). Furthermore, this HPT was also operated at -20°C and NEP was reduced to 1.83×10^{-14} W/Hz$^{1/2}$ at the same operating conditions.

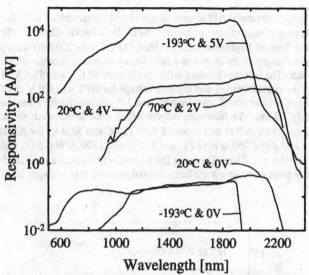

Figure 1. Responsivity variation with wavelength at different bias voltages and temperatures for an HPT sample (A1-d2) [11].

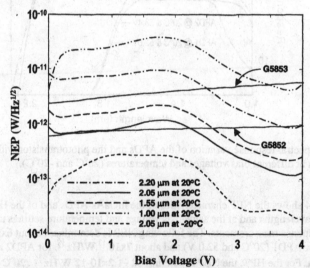

Figure 2. Comparison of the NEP variation with bias voltage between HPT and commercial p-i-n photodiodes [10]. The NEP of the p-i-n photodiodes were measured at 2.0 μm, 1.0 V and -20°C.

In addition, commercial APDs were acquired and characterized at NASA LaRC. Spectral response measurements of InGaAs APDs from Perkin-Elmer (APD1 with 80 μm diameter) and Sensors Unlimited Inc./Goodrich (APD2 with 200 μm) were performed. Figure 3 shows responsivity curves as a function of wavelength of the InGaAs APDs and phototransistor. The selected biases for the APDs were 52.0 V and 53.4 V at 20°C as specified by the manufacturers and the bias voltage for HPT was 4.0 V at -20°C. These two APDs have similar cutoff wavelengths near 1.7 μm, but the HPT has the cutoff at approximately 2.1 μm. The following responsivities were determined: above 100 A/W (20°C and 53.4 V) for APD2 and about 85 A/W (20°C and 52.0 V) for APD1 at 1.6 μm wavelength; and about 260A/W (20°C and 3.0 V) and 1700 A/W (-20°C and 4.0 V) for the phototransistor at 2.0 μm. These are the highest responsivities measured for those APDs and the phototransistor at different breakdown and bias voltages at 20°C and -20°C, respectively.

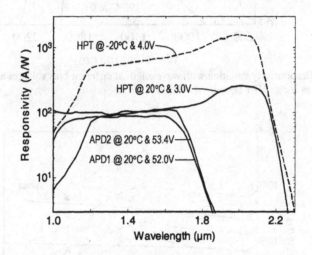

Figure 3. Spectral response variation of the APDs and the phototransistor with wavelength at different bias voltages and temperatures (20°C and -20°C).

Figure 4 shows the NEP characteristics of the InGaAs APDs and of the HPT at different wavelengths and at the same bias voltage and temperature settings used for the spectral response measurements. The following NEP is determined: about 6×10^{-15} W/Hz$^{1/2}$ for APD1 (20°C and 52.0 V) and about 7×10^{-13} W/Hz$^{1/2}$ for APD2 at 1.6 μm wavelength. For the HPT, the NEP is determined $\sim 1.2 \times 10\text{-}12$ W/Hz$^{1/2}$ (20°C and 3.0 V) and $\sim 4.6 \times 10^{-14}$ W/Hz$^{1/2}$ (-20°C and 4.0 V) at 2.0 μm wavelength. Comparison with the tested avalanche photodiodes measured at 20°C shows that at breakdown voltage, APD1 achieves near two-order-of-magnitude lower NEP than APD2 and one-order-of-magnitude lower NEP than phototransistor, when it was cooled down to -20°C. We believe that these discrepancies are due to the higher noise of APD2 and the phototransistor as compared to APD1.

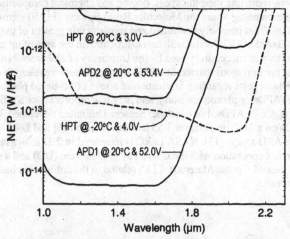

Figure 4. Comparison of NEP between the APDs and the phototransistor at the same operating conditions (20°C and -20°C).

The HPTs have longer wavelength sensitivity and comparably short wavelength sensitivity, which is an attractive solution for the broadband imaging systems. The attraction to this solution is because the easiest way to acquire broadband imaging is to use a single detector array with simple optics and electronics. For commercial technologies (for example, Si-cutoff 1.1 µm and InGaAs: 1.7 or extended 2.6 µm detectors), each wavelength band requires a separate detector with appropriate dichroic, focusing optics, support electronics, cooling, and mounting hardware. These requirements for additional components increase the size of an imaging system, resulting in increased power consumption, weight, and cost. Therefore, the broadband approach is being pursued because it offers an advantage of simple, efficient, cost-effective, lightweight, and compact instrument design. Broadband phototransistor arrays proposed to the NASA Earth-Sun Systems Division (ESSD) technology program for the active and passive remote sensing instrument will operate in the spectral range of 0.6 – 2.5 µm and extend this technology to flight missions for the next generation. For the broad-spectral active and passive instrument, we will select regions that cover only interest lines (e.g., CO_2, CH_4, H_2O, CO) with higher resolution within those bands.

FUTURE DEVELOPMENT OF NEAR-INFRARED DETECTOR FPA

The record results achieved with discrete 200-μm diameter HPTs provide a base for designing and fabrication of arrays of these devices. However, fabrication of the arrays is a challenging task. It requires a very high level of homogeneity of InGaAsSb/AlGaAsSb epitaxial layers, including their thickness, doping and chemical composition. For this purpose we are planning to use the Molecular Beam Epitaxy (MBE) - growth method. This method provides precisely controlled and reproducible growth of the required multilayer InGaAsSb/AlGaAsSb/GaSb heterostructure on the area of up to 4 cm^2. This epitaxial growth was successfully used by the University of Delaware-NASA team for fabrication of the first room temperature HPT with the cutoff wavelength of 2.4 μm [15].

There are no reports regarding fabrication of arrays of Sb-based photodetectors with internal gain (APDs or phototransistors), and only few reports about SWIR arrays of InGaAs and HgCdTe APDs. For example; Sensors Unlimited developed 32 x 32 arrays of InGaAs APDs on a 100 μm pitch for 1.55 μm wavelength [16] and DRS reported about first HgCdTe APD arrays [17]. NASA LaRC is interested in 2-D arrays fabrication and hybridization in cooperation with the University of Delaware (UD) and a start-up company Advanced Optical Materials, LLC, related to the antimonide based phototransistors.

SUMMARY

We have demonstrated current single-element infrared detectors and future development of near-infrared focal plane arrays (FPA) for applications to next generation space-based instruments. Most of these activities are based on phototransistor technology, which is offering high internal gain and low NEP operation. Developed phototransistors showed higher responsivity and higher gain with a light collecting area diameter of 200 μm. Spectral response measurements and NEP calculations were carried out to the phototransistor to achieve the responsivity at 2.0 μm and NEP is several times lower than commercially available state-of-the-art p-i-n photodiodes. On the other hand, phototransistor has achieved two-order-of-magnitude higher responsivity as compare to APDs, despite the higher NEP as compare to Perkin-Elmer APD and lower NEP with respect to SUI/Goodrich APD at near room temperature operation. These new phototransistors will improve the sensitivity of the passive remote sensing instruments and reduce the laser transmitter power for active remote sensing instruments, while eliminating the need for high breakdown voltage avalanche photodiode.

REFERENCES

1. P. Ambrico, A. Amodeo, P. Girolamo, and N. Spinelli, Applied Optics 39(36), 6847–6865 (2000).
2. S. Ismail, G. J. Koch, B. W. Barnes, N. Abedin, T. F. Refaat, J. Yu, S. A. Vay, S. A. Kooi, E. V. Browell, U. N. Singh, Proceedings of the 22nd International Laser Radar Conference, 65-68 (2004).

3. T.F. Refaat, M.N. Abedin, O.V. Sulima, S. Ismail, and U.N. Singh, Optical Engineering 43(7), 1647-1650 (2004).
4. T.F. Refaat, M.N. Abedin, O.V. Sulima, U.N. Singh, and S. Ismail, IEDM Tech. Dig., 355-358 (2004).
5. V. Diadiuk, S.H. Groves, C.E. Hurwitz, and G.W. Iseler, IEEE J. Quantum Electronics, QE-17 (2), 260-264 (1981).
6. J.C. Campbell, A.G. Dentai, W.S. Holder and B.L. Kasper, Electron. Lett. 19, 818-820 (1983).
7. K.K. Loi and M. Itzler, Compound Semiconductor 6(3), 1-3 (2000).
8. O.V. Sulima, T.F. Refaat, M.G. Mauk, J.A. Cox, J. Li, S.K. Lohokare, M.N. Abedin, U.N. Singh, and J.A. Rand, Electronics Letters 40, 766-767, (2004).
9. O.V. Sulima, T.F. Refaat, M.G. Mauk, J.A. Cox, J. Li, S.K. Lohokare, M.N. Abedin, U.N. Singh, and J.A. Rand, Presented at 6th Middle-Infrared Optoelectronics Materials and Devices (MIOMD) Conference, in St. Petersburg, Russia, 28 June-1 July 2004.
10. M.N. Abedin, T.F. Refaat, O.V. Sulima, and U.N. Singh, IEEE Trans. Electron Devices 51(12), 2013-2018 (2004).
11. M.N. Abedin, T.F. Refaat, O.V. Sulima, and U.N. Singh, International Journal of High Speed Electronics and Systems, v.15, No.2, pp. 567-582, (2006).
12. T.F. Refaat, M.N. Abedin, O.V. Sulima, S. Ismail, and U.N. Singh, Optical Engineering, v. 43, No. 7, pp.1647-1650 (2004).
13. T.F. Refaat, M.N. Abedin, O.V. Sulima, U.N. Singh and S. Ismail, Technical Digest of the 50th IEEE International Electron Devices Meeting (IEDM), San-Francisco, Ca, December 2004, pp.355 – 358 (2004).
14. T.F. Refaat, M. N. Abedin, O.V. Sulima, S. Ismail and U.N. Singh, Proc. SPIE, v.5887, pp.588706-1 – 588706-13 (2005).
15. O.V. Sulima, K. Swaminathan, T.F. Refaat, N.N. Faleev, A.N. Semenov, V.A. Solov'ev, S.V. Ivanov, M.N. Abedin, U.N. Singh, and D. Prather, Electronics Letters, v.42 (1), pp. 55-56 (2006).
16. J.C. Dries, T. Martin, W. Huang, M.J. Lange, and M. J. Cohen, IEEE LEOS Proceedings, 2002.
17. J.D. Beck, C.-F. Wan, M.A. Kinch, and J.E. Robinson, SPIE v. 4454, pp. 188-197 (2001).

Mater. Res. Soc. Symp. Proc. Vol. 1076 © 2008 Materials Research Society   1076-K03-07

Structural and photoconducting properties of MBE and MOCVD grown III-nitride double-heterostructures

Sindy Hauguth-Frank[1], Vadim Lebedev[2], Katja Tonisch[1], Henry Romanus[1], Thomas Kups[1], Hans-Joachim Büchner[3], Gerd Jäger[3], Oliver Ambacher[2], and Andreas Schober[1]

[1]Institute of Micro- and Nanotechnologies, Ilmenau Technical University, Gustav- Kirchhoff-Str. 7, Ilmenau, 98693, Germany

[2]Fraunhofer Institute for Applied Solide State Physics, Tullastr. 72, Freiburg, 79108, Germany

[3]Institute of Measurement and Sensor Technology, Ilmenau Technical University, Gustav-Kirchhoff-Str. 7, Ilmenau, 98693, Germany

ABSTRACT

Investigations on standing wave (SW) interferometry come in focus of interest in the course of ongoing miniaturization of high precision length measurement systems. A key problem within these efforts is the development of a transparent ultra-thin photodetector for sampling the intensity profile of the generated SW. Group III-materials are promising candidates to ensure a good photodetector performance combined with the required optical transparency. In this work, we report on the interrelation of strain and dislocation density along with the influence of the structural properties on the sensitivity of double-heterostructure III-nitride photodetectors grown by molecular beam and metal organic vapour phase epitaxy.

INTRODUCTION

Optical interferometry has become the most used method for nanoscale high precision displacement measurements, especially, in areas where a non-contact measurement setup is required. State of the art interferometers are based on the Michelson interferometer concept. These interferometers consist of a plurality of classical optical components (e.g. beam splitters), which require a relatively large workspace along with high precision alignment. The most promising alternative is the standing wave interferometer (SWI), which combines the advantages of classical interferometers with the moderate space requirements due to a reduced number of the optical components [1]. In the fundamental setup of SWI most of the incident beam is transmitted through the photodetector, travels to the mirror on a moving object and is reflected to the photodetector again. Thus, the key element of SWI is the nearly transparent photo-sensitive element, which has to sample the intensity of the generated standing electro-magnetic wave to recognize the movement of the object associated with the change of the position of the standing wave extremes. A novel approach in the development of such a photosensitive element is the usage of III-nitride-based double-heterostructure photodetectors (DHPs). The main advantages for building up a photodetector with III-nitrides arise from the large bandgaps and associated optical transparency in the visible part of the spectrum along with high temperature stability [2].

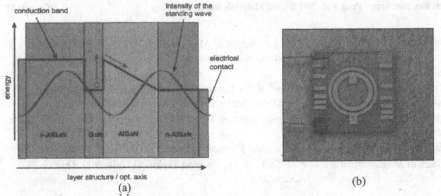

(a) (b)

Figure 1. Concept of III-nitride based DHP for SWI (b) DHP with external interconections

The photodetector designed for a laser wavelength of 633nm (He-Ne line) consists of a $Al_xGa_{1-x}N/GaN$ double-heterostructure and a ~20 nm thick GaN active layer, which acts as a potential well for free electrons (figure 1). Barriers of ~0.5 eV to 1.83 eV are built in the structures due to the band offset between GaN and $Al_xGa_{1-x}N$ [3]. The excitation of the electrons accumulated within the GaN potential well is forced by the absorption of laser light. It allows the carriers to overcome the barrier, and then the photocurrent can be quantified. To compensate polarization effects at the AlGaN/GaN and GaN/AlGaN interfaces and to offer additional electrons for the GaN potential well, the n-AlGaN cathode layer has been Si-doped (up to N_{Si} ~ $\cdot 10^{18}$ cm^{-3}).

As the optical and photoconducting properties of DHPs are mainly affected by the crystal quality of the epilayers, the issues related to the structural properties and their impact on the resulting sensitivity of DHPs will be discussed in this work. In particular, the interrelation of strain and dislocation density in DHPs grown by molecular beam epitaxy (MBE) and metal organic chemical vapour deposition (MOCVD) is reported.

EXPERIMENT

The first set of DHP samples were grown in a Balzer's plasma induced molecular beam epitaxy (PIMBE) system. The growth of the second set of DHPs was accomplished using a AIX200RF MOCVD system. All DHP samples were grown on double side polished (0001) on-axis sapphire substrates. The detailed description of PIMBE and MOCVD equipment and characterization methods used can be found elsewhere [4, 5]. Due to the differences in the growth techniques, the layer thickness has been chosen to be different for PIMBE and MOCVD samples. The MBE samples consist of a 160 nm thick AlN nucleation layer followed by a 120 nm thick $Al_{0.35}Ga_{0.65}N$ layer grown to relief the enormous strain induced by the mismatch between the sapphire substrate and the III-nitride materials. The core DHP MBE structure deposited on the first AlGaN layer consists of a 80 nm thick Si doped $Al_{0.35}Ga_{0.65}N$ layer, which forms the cathode, a 20 nm thick GaN potential well and a 60 nm thick $Al_{0.35}Ga_{0.65}N$ barrier layer. The MOCVD structures have 20 nm and 1.1 µm thick AlN nucleation and GaN buffer layers, respectively. The core structure of the MOCVD samples consists of a 35 nm thick Si doped $Al_{0.33}Ga_{0.67}N$ cathode layer, a 20 nm thick GaN active layer and a 35 nm thick $Al_{0.5}Ga_{0.5}N$ barrier. In order to avoid cracks within the DHP core [6], the thickness of $Al_{0.5}Ga_{0.5}N$ layer was drastically reduced. To ensure good electrical contacts, a 4 nm thick Si doped AlGaN layer was deposited on top of all DHPs. Finally, the grown heterostructures have been analysed by a high-resolution x-ray diffraction (HRXRD) and a cross-sectional transmission electron microscopy (X-TEM).

For the fabrication of DHP, the AlGaN/GaN structures were etched isotropically by means of a chlorine based inductive coupled plasma process (ICP) using a 70 nm thick Ni etch mask. The etched mesa structure has been contacted using sputtered 20 nm Ti / 150 nm Au pads. The sensitivity dependencies of DHP were measured by photo-assisted voltage-current (I-U) measurements using a lock-in amplifier technique to receive the frequency resolved data.

RESULTS AND DISCUSSION

Information on strain, composition and orientation of the epilayers was aquired by recipical space maps of the (-105) reflex and the (002) reflex of the AlGaN/GaN DHPs using HRXRD.

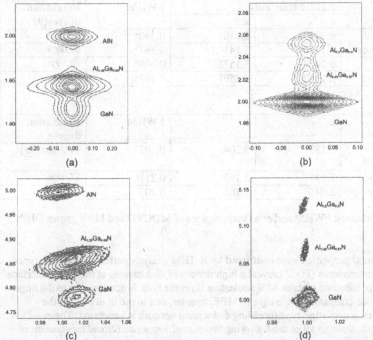

Figure 2. Reciprocal space maps for the symmetric (002) and asymmetric (105) X-ray reflection peaks (displayed with the Miller indices at all axes) of DHP samples grown by MBE (a and c) and MOCVD (b and d)

As shown in figure 2 for MBE samples, the AlN layer is almost relaxed (table 1) on the sapphire substrate, while the succeeding AlGaN and GaN epilayer grow pseudomorphically on the AlN buffer. Both layers indicate a slight compressive strain as shown by the lattice constants in table 1. The AlGaN barrier grown on top of the GaN is compressively strained as well and shows the highest strain within the structure with a degree of relaxation of only ~27%.

In contrast to the MBE samples, HRXRD on the MOCVD grown AlGaN/GaN DHPs reveals a relaxed GaN buffer on top of the AlN nucleation layer along with the pseudomorphical growth of the following AlGaN and GaN epilayers. The lattice constant analysis for the MOCVD samples reveals also a high tensile strain in the cathode and barrier AlGaN layers. In this case, the GaN intermediate layer reduces a high amount of the tensile strain (R = 91 %) and is therefore relaxed. The origin of the higher strain in the MOCVD samples is shown by the FWHM of the different layers of both structures (table 1). The smaller FWHM value of the MOCVD layers expose a lower density of slightly misoriented crystallites within the material and in conclusion a lower density of edge dislocations [7,8]. Therefore the stress induced by the difference in the lattice constants of the epilayers, can not be reduced.

MBE				
Layer	Lattice constants		FWHM	Relaxation degree[9]
	c	a		
AlN	4.9799	3.1108	0.340°	100%
$Al_{0.35}Ga_{0.65}N$ cathode	5.1276	3.1410	0.367°	57,05%
GaN	5.2038	3.1575	0.409°	73,19%
$Al_{0.35}Ga_{0.65}N$ barrier	5.1152	3.1591	0.367°	27,31%
MOCVD				
Layer	c	a	FWHM	Relaxation degree
GaN buffer/ intermediate	5.185	3.204	0.197°	91% (buffer)
$Al_{0.33}Ga_{0.67}N$ cathode	5.111	3.197	0.211°	17,69%
$Al_{0.5}Ga_{0.5}N$ barrier	5.046	3.188	0.207°	29,76%

Table 1. Lattice constants, FWHM and relaxation degrees of MOCVD and MBE grown DHPs.

These structural properties were confirmed by X-TEM analysis performed on both types of DHPs. These measurements (Fig. 3) reveal a high density of dislocations at the heterointerface between the sapphire substrate and the AlN nucleation layer for both structures, due to the huge mismatch in the lattice constants. In the case of MBE samples, due to the Si doping of the cathode AlGaN layer, a new edge-type threading dislocation network is generated. These dislocations propagate through all of the following layers enabling an additional mechanism of strain relaxation.

In the MOCVD samples the AlN nucleation layer is only 20 nm thick. It grows three dimensionally forming "islands" having a small grain size, and therefore, relieving a high amount of strain. This nucleation layer is followed by a 1.1 μm thick GaN buffer, which reduces the dislocation density down to $2*10^9$ cm^{-1} at the transition to the AlGaN cathode layer due to the favorable growth conditions. At the start of the AlGaN and the succeeding GaN intermediate and AlGaN barrier layers only less additional dislocations are initiated in despite of a high strain. Therefore the structural quality of the core DHP layer remains high.

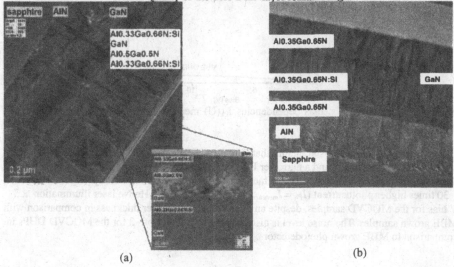

(a)   (b)

Figure 3. X-TEM of MOCVD (a) and MBE (b) grown DHP structures

The structural quality of the grown layers affects the sensitivity of the DHP. Especially, the dislocations appear as traps, which slow down or terminate photoelectrons excited by the incoming laser light. In terms of sensitivity, the quality of the grown epilayers in the DHP core structure should be at a maximum enhancing the signal-to-noise ratio [10]. Additionally, the density of dislocations in the AlGaN barrier layer has a direct impact on the level of the dark current of the DHP, and therefore, on the usable signal for analyzing the intensity of the standing wave profile [11].

159

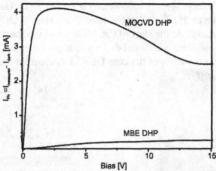

Figure 4. Photocurrent vs. bias dependencies $I_{Ph}(U)$ measured on MOCVD and MBE grown DHPs

Due to the higher crystal quality that is proven by HRXRD and TEM measurements (e.g. a lower dislocation density and a narrower FWHM), the MOCVD samples show also a better detection performance. In particular, the measurements of the photosensitivity (figure 4) reveal a ~ 30 times higher photocurrent ($I_{Ph} = I_{measured} - I_{dark}$) under 5 mW He-Ne laser illumination at 3 V bias for the MOCVD samples, despite an over 40% lower barrier thickness in comparison with MBE grown samples. The noise level is also reduced by a factor of ~3 for the MOCVD DHPs in comparison to MBE grown photodetectors.

CONCLUSIONS

In conclusion, we reported on the structural properties and photo-sensitivity of III-nitride DHPs designed for standing wave interferometry at $\lambda = 632$ nm. It is shown that the crystal quality of DHP structures is highly dependent on both the heterostructure design and the growth technique used. The MBE grown samples demonstrate an almost relaxed crystal structure accompanied by the formation of a high density threading dislocation network negatively affecting the photodetector performance. In contrary, the MOCVD grown samples are characteristic by a reduced density of dislocations accompanied by an enormous built-in strain in the core DHP structure. It is shown that despite the opportunity of growing thicker barrier layers by MBE, a higher photocurrent / dark current ratio and a low noise level can be reached on MOCVD structures.

ACKNOWLEDGMENTS

This work was supported by the Arbeitsgemeinschaft industrielle Forschungs-vereinigungen (AIF) within the project "Entwicklung eines transparenten Quantum-Well-Fotosensors für ein Stehende-Wellen-Interferometer", AiF-Vorhaben 14634 BR.

REFERENCES

1. H- J Büchner, H Stiebig, V Mandryka, E Bunte, G Jäger, Meas. Sci. Technol. **14**, 311-316 (2003).
2. O. Ambacher, J. Phys. D: Appl. Phys. **31**, 2653 (1998).
3. I. Vurgaftman and J. R. Meyer, J. Appl. Phys. **94**, 3675 (2003).
4. V. Lebedev, V. Cimalla, T. Baumann and O. Ambacher, F. M. Morales, J. G. Lozano and D. González, J. Appl. Phys. **100**, 094903 (2006).
5. V. Lebedev *et al.* J. Appl. Phys. **101**, 054906 (2007).
6. W. H. Sun, J. P. Zhang, J. W. Yang, H. P. Maruska, M. Asif Khan, R. Liu, and F. A. Ponce, Appl. Phys. Lett. **87**, 211915 (2005).
7. T. Ive: Growth and investigation of AlN/GaN and (Al,In)N/GaN based Bragg reflectors; Humboldt-Universität Berlin 2005.
8. M. Hermann, F. Furtmayr, A. Bergmair et al.; Appl. Phys. Lett **86** 192108 (2005).
9. M. Schuster *et al.* J. Phys. D: Appl. Phys. **32** A56-A60 (1999).
10. V. Lebedev, F. M. Morales, H. Romanus *et al.* J. Appl. Phys. **98** 093508 (2005).
11. S.C. Jain, W. Willander, J. Narayan *et al.* J. Appl. Phys. **87** 965-1006 (2005).

Nanocrystal and Photonic Structure Devices

Mater. Res. Soc. Symp. Proc. Vol. 1076 © 2008 Materials Research Society   1076-K05-02

A Buried Silicon Nanocrystals Based High Gain Coefficient SiO2/SiOX/SiO2 Strip-Loaded Waveguide Amplifier on Quartz Substrate

Cheng-Wei Lian, and Gong-Ru Lin

Graduate Institute of Photonics and Optoelectronics and Department of Electrical Engineering, National Taiwan University, No. 1, Roosevelt Rd. Sec. 4, Taipei, 10617, Taiwan

ABSTRACT

A Si-rich SiO_X strip-loaded waveguide with silicon (Si) nanocrystal contributed amplified spontaneous emission at 750-850 nm with the associated spectral linewidth of 140 nm is characterized. By using the variable stripe length (VSL) method we demonstrate the optical gain and loss coefficients of 65 and 5 cm^{-1}, respectively, for such a waveguide amplifier. The optical gain and loss coefficients are observed by fitting the one dimensional amplifier equation. The small-signal power gain of 18.4 dB at the wavelength of 805 nm under He-Cd laser pumping of 40 mW at 325 nm is obtained from the $SiO_2/SiO_X/SiO_2$ waveguide amplifier with a length of 1 cm.

INTRODUCTION

Prior to the development of nano-synthesis technology, Si was never considered as a candidate for light emitters owing to its indirect electronic bandgap characteristics. Over the last two decades, versatile low-dimensional Si structures including nanocrystallite Si [1-2], porous Si [3], Si/insulator superlattices, and Si nano-pillars etc. with intensive luminescence and optical gain have been demonstrated [1]. The first observation on optical gain of silicon nanocrystal was proposed by Pavesi and co-workers, while the net modal gain coefficient of the silica waveguide with buried Si nanocrystal synthesized by Si^{+}-ion implantation was determined as 100 cm^{-1} [1]. Later on, Luterova and co-workers also observed the optical gain and loss coefficient of 25 and 15 cm^{-1} [3], respectively, for the porous SiO_2 nanograins embedded in SiO_2 deposited by the sol-gel method. However, the gain of Si nanocrystal synthesized within the SiO_X film grown by the PECVD method was seldom addressed, in which the density of Si nanocrystal is reported to be the highest among all mentioned samples. The concept and fabrication of all Si-based photonic integrated circuitry become achievable on such a mainstay semiconductor in microelectronic circuitry by employing Si based materials as alternative candidates for light emission. The aforementioned Si structure in a low-dimensional form, or the Si substrate with selected active impurities (such as erbium) inserted into Si lattice [4-5], or new composite phases (such as iron disilicide). In principle, the physical mechanism underlying the high external quantum efficiencies for photoluminescence in low-dimensional Si is mainly attributed to the quantum confinement of excitons in the nano-scale crystalline Si matrix. With the rapid evolution on the plasma enhanced chemical vapor deposition (PECVD) grown Si-rich SiO_X film containing dense Si nanocrystal, the development of low-dimensional Si nanocrystal based near-infrared luminescent materials and light emitting diodes have recived much attention during the past decade. More recently, many claims on the future role for Si in photonic applications with versatile optical device configurations have emerged, such as waveguides [2], ring resonators [6], ring disks filters [7], arrayed-waveguide gratings [8], etc. However, the Si based waveguide amplifiers and lasers have not yet been completely realized, still remaining an uncertain issue toward the all-Si-based photonic integrated circuitry. Particularly, the SiO₂ strip-loaded waveguide structure which exhibits better optical mode confinement than the conventional planar waveguide geometry has never been employed to investigate the SiO$_X$ with buried Si nanocrystal. In this work, we design and fabricate the strip-loaded waveguide with buried Si nanocrystal and determine the optical gain coefficient of Si nanocrystal. The Si-rich SiO$_X$ film was grown upon the quartz substrate by using PECVD and annealed in a quartz furnace with flowing N₂ to precipitate Si nanocrystal. Afterwards, we will discuss the amplified spontaneous emission (ASE) spectrum of the Si-rich SiO$_X$ strip-loaded waveguide under He-Cd laser pumping of 40 mW with a wavelength of 325 nm. The VSL method is used to observe the optical net model gain and loss coefficient by fitting to the one dimensional amplifier equation. Then, we will demonstrate the small-signal amplification in the Si nanocrystal strip-loaded waveguide. The Si nanocrystal has such high optical gain coefficient, and the Si substrate based characteristic, that these findings could lead the way to a Si based laser.

EXPERIMENT

First of all, the single-mode strip-loaded waveguide is simulated to observe the configuration parameters for the SiO₂/SiO$_X$:nc-Si/SiO₂. The tri-layer planar waveguide structure for the effective index method (EIM) calculation is shown in figure 1.

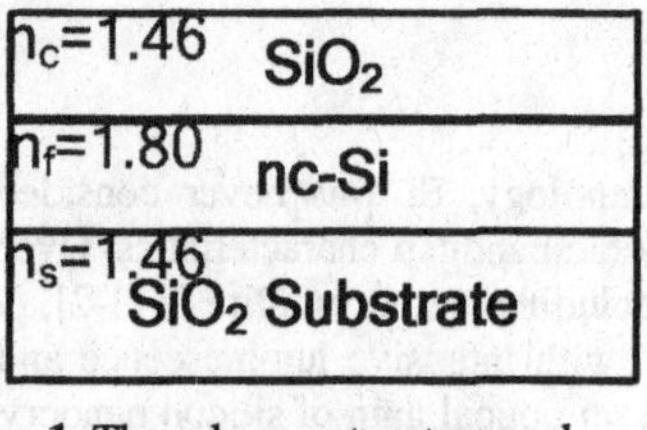

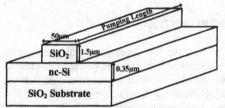

| n_c=1.46 SiO₂ |
| n_f=1.80 nc-Si |
| n_s=1.46 SiO₂ Substrate |

Figure 1. Three layers structure and definitions for EIM calculation.

Figure 2. The quartz based strip-loaded waveguide structure.

The EIM is employed to calculate the thickness of active SiO$_X$ layer for single mode confinement. The EIM equations are shown as below,

$$
\begin{cases}
V = kh\sqrt{n_f^2 - n_s^2} \\
b = \dfrac{N^2 - n_s^2}{n_f^2 - n_s^2} \\
N = \dfrac{\beta}{k} = n_f \sin\theta
\end{cases}
\quad
\begin{cases}
a_{TE} = \dfrac{n_s^2 - n_c^2}{n_f^2 - n_s^2} \\
a_{TM} = \dfrac{n_f^4}{n_c^4}\dfrac{n_s^2 - n_c^2}{n_f^2 - n_c^2} = \dfrac{n_f^4}{n_c^4} \times a_{TE} \\
V = \dfrac{m\pi + \tan^{-1}\sqrt{\dfrac{b}{1-b}} + \tan^{-1}\sqrt{\dfrac{b+a}{1-b}}}{\sqrt{1-b}}
\end{cases}
\tag{1}
$$

The parameter V is the normalized film thickness, b is the normalized guide index and N is the effective index. In our case, $n_{nc\text{-}Si}$ = 1.8, n_{SiO2} = 1.46 at the wavelength of 800 nm corresponding to a photoluminescence (PL) peak. We calculated that the thickness of SiO$_X$:nc-Si layer for single-mode confinement can be widely tunable from 0 to 380 nm. Afterwards, the strip-loaded waveguide structure shown in figure 2 was determined except for the design of the width and height of SiO$_2$ strip. The width of strip must be sufficiently large, on the order of 50 μm, to ensure that the strip pattern will not be over-etched during the wet-etching process. The simulation software, Beam Propagation Method (BPM), supported by R-Soft Company is employed to confirm the height of the strip and to simulate the output mode field, which helps us to achieve optimized optical confinement within the waveguide structure.

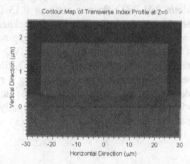

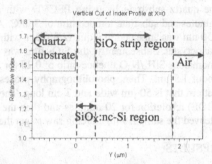

Figure 3. The contour map of the transverse index profile.

Figure 4. The cross-sectional view of the index profile.

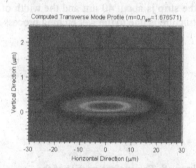

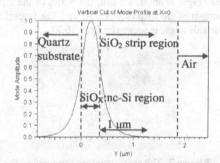

Figure 5. The simulation of the contour map of the fundamental mode confinement.

Figure 6. The cross-sectional view of the simulation on the confined fundamental mode.

Figure 3 shows the contour mapping of the transverse index profile, the width and height of the SiO$_2$ strip is 50 μm and 1.5 μm, respectively, and the thickness of SiO$_X$:nc-Si layer is 350 nm. If the SiO$_X$:nc-Si layer is thicker there is multi-mode confinement. The height of the strip region has a minimum value, if the height is smaller than the minimum value, the optical power leaks to the air region. The width of the strip region has a maximum value, if the width is larger than the maximum value, it becomes multi-mode. Figure 4 shows the cross-sectional view of the index profile, in which the refractive index of SiO$_2$ is 1.46, the refractive index of SiO$_X$ is 1.80 and the refractive index of air region is 1.00. Figure 5 shows the simulation of contour mapping of fundamental mode confinement, it is clearly shown that the fundamental mode can be entirely confined in the SiO_X:nc-Si layer. Figure 6 shows the cross-sectional view of the simulation on the confined fundamental mode, the power of the fundamental mode vanishes at heights of 1.3 μm above the quartz substrate, and the height of the SiO_2 strip needs to be around 1 μm. For larger tolerance, we chose the height of the SiO_2 strip to be 1.5 μm. The optical power is confined in the SiO_X:nc-Si region when the optical signal and ASE are confined in the active region.

In the experiment, we fabricated a single-mode strip-loaded waveguide with buried Si nanocrystal on a 2 cm-square quartz substrate. First of all, the Si-rich SiOx film was grown upon the quartz substrate by using PECVD with anomalous SiH_4/N_2O fluence ratio of 0.15 for 8 minutes and substrate temperature of 350°C. After deposition, the SiOx film with a thickness of 350 nm was annealed in a quartz furnace with flowing N_2 at 1100°C for 90 min to precipitate Si nanocrystal. Subsequently, the SiO_2 film was grown upon the SiOx film by using PECVD with anomalous SiH_4/N_2O fluence ratio of 0.25 for 20 minutes yielding a thickness of the SiO_2 film of about 1.5 μm. Then, photolithography processes were performed on the sample using a mask pattern that is 50 μm wide and 3 cm long. The sample was wet etched by buffered oxide etching (BOE) resolution for 20 minutes and then the photo resist was removed by acetone. Finally, we cleaved the sample with a dicing saw to get the input and output ends.

RESULTS

Figure 7 shows the top view of the strip structure sample imaged by an optical microscope. Figure 8 shows the surface profile of the strip-loaded waveguide sample measured by alpha step, the height is about 1.5 μm, the width of the top of the strip is about 40 μm and the width of the bottom of the strip is about 50 μm. The trapezium structure is due to the wet etching process and the side angle is approximately 16.7°.

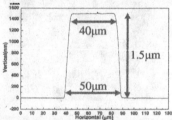

Figure 7. The strip structure top view of the sample imaged by optical microscope.

Figure 8. The surface profile of the strip-loaded structure measured by alpha step.

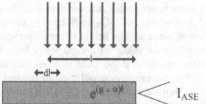

Figure 9. One dimensional amplifier model.

We use the Variable Stripe Length (VSL) method to obtain the gain coefficient of the $SiO_2/SiO_x/SiO_2$ strip-loaded waveguide on the quartz substrate. The ASE intensity (I_{ASE}) that is emitted from the sample edge is measured as a function of the pumping length (l). From a fit of the resulting curve, the optical gain coefficient (g) and loss coefficient (α) can be deduced at every wavelength. By assuming a one dimensional amplifier model as shown in figure 9, I_{ASE} can be related to g and α by the following formulas (2) where I_{spon} is spontaneous emission intensity.

$$dI_{ASE} = I_{spon} \times dl \times e^{(g-\alpha)l}$$
$$\Rightarrow \int dI_{ASE} = \int I_{spon} \times dl \times e^{(g-\alpha)l}$$
$$\Rightarrow I_{ASE} = \frac{I_{spon}}{g-\alpha} \times e^{(g-\alpha)l} + c$$
$$when \quad l = 0, I_{ASE} = 0 \qquad \qquad (2)$$
$$\Rightarrow 0 = \frac{I_{spon}}{g-\alpha} + c$$
$$\Rightarrow c = -\frac{I_{spon}}{g-\alpha}$$
$$\Rightarrow I_{ASE} = \frac{I_{spon}}{g-\alpha} \times (e^{(g-\alpha)l} - 1)$$

Figure 10 shows the experimental setup of the VSL method. The Si nanocrystal strip-loaded waveguide is top-illuminated by a He-Cd laser at 325 nm with 40 mW power. The Amplified Spontaneous Emission (ASE) intensity is received at the waveguide edge by a fiber with a 400 μm core and is then characterized by an optical spectrum analyzer. By moving the slit, the pumping length is changed, and we obtain the different ASE intensities. Figure 11 shows the ASE spectrum gathered by the VSL method. The ASE spectrum ranges between 750-850 nm with the central wavelength of 805 nm and the 3 dB spectral linewidth of 140 nm. Figure 12 is the complete experimental setup. We used a UV lens to expand the laser beam and a UV mirror to reflect the laser beam on the top to pump the waveguide. The UV cylindrical lens can focus the laser beam in one dimension. We collect the ASE spectrum with the fiber having a core of 400μm. Figure 13 shows a photograph of the waveguide which is pumped by the He-Cd laser.

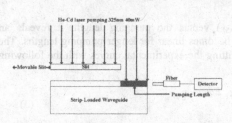

Figure 10. The VSL experimental setup.

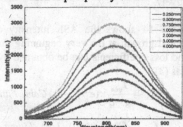

Figure 11. The ASE spectrum gathered by the VSL method.

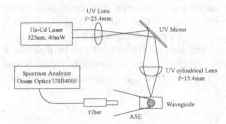

Figure 12. The total experimental setup.

Figure 13. A photograph of the waveguide which is pumped by the He-Cd laser.

DISCUSSION

The broadband ASE spectrum with high optical gain coefficient is observed by characterizing the waveguide with the VSL method. Under the pumping wavelength of 325 nm and power of 40 mW, a net model gain coefficient of up to 70 cm⁻¹ contributed by the Si nanocrystal is reported. The gain coefficient is fitting to the waveguide length dependent ASE intensity, corresponding to a power gain of 18.4 dB at the wavelength of 805 nm.

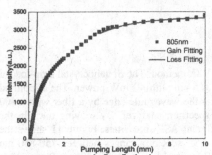

Figure 14. The ASE intensity at wavelength of 805 nm versus He-Cd laser pumping length. And the fitting line of gain and loss of waveguide, it is 65 cm⁻¹ and 5 cm⁻¹ respectively.

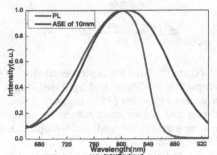

Figure 15. The normalized PL spectrum and ASE spectrum.

Figure 14 shows the ASE intensity (I_{ASE}) versus the pumping length. It reveals an exponential growth at the very beginning that becomes linear for longer pumping lengths. The gain and loss coefficients can be obtained by fitting the experimental curves with the following formulas (3),

$$\begin{cases} I_{ASE} = \dfrac{I_{spon}}{g-\alpha} \times (e^{(g-\alpha)l} - 1) \;\; Gain \;\; Fitting \;\; Equation \\[3mm] I_{ASE} = \dfrac{I_{spon}}{-\alpha} \times (e^{-\alpha l} - 1) \;\; Loss \;\; Fitting \;\; Equation \end{cases} \tag{3}$$

By fitting the VSL exponential growth curve within the standard one-dimensional amplifier model, the exponentially grown ASE intensity with waveguide length is attributed to the net model gain coefficient of up to 65 cm^{-1}. The linear I_{ASE} curve at longer pumping length region occurs due to the gain saturation effect, obtained the Si nanocrystal strip-loaded waveguide loss coefficient of 5 cm^{-1} from experiments. The large loss coefficient is because of Si nanocrystal scattering. This results in the total gain coefficient of up to 70 cm^{-1}. Such a high gain coefficient obtained from Si nanocrystal based waveguide is premier and is attributed to the optimization of Si-rich SiO$_X$ film via PECVD growth. The small-signal power gain of 18.4 dB at the wavelength of 805 nm under He-Cd laser pumping of 40 mW at 325 nm is obtained from the SiO$_2$/SiO$_X$/SiO$_2$ waveguide amplifier with a length of 1 cm.

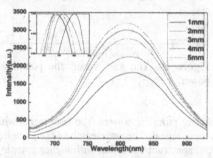
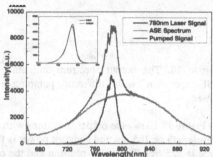

Figure 16. The ASE spectrum with different pumping lengths. The inset is the normalized ASE spectrum.

Figure 17. The small signal pumped spectrum.

Figure 15 shows the normalized photoluminescence (PL) spectrum of Si nanocrystal and the ASE spectrum, the peak wavelength is red shifted about 6 nm from the non-guiding structure to the guiding structure. It is because of the waveguide structure that the luminescent wave is guided in the strip-loaded waveguide; only the specific mode can be confined in the waveguide. The wavelength of the ASE guided mode is 6 nm different to the spontaneous emission. Figure 16 shows the normalized ASE spectrum of Si nanocrystal with different pumping length. The peak wavelength of ASE spectrum is blue shifted with longer pumping lengths. Because of the longer pumping length, the mode guiding is stronger and the peak wavelength becomes stable. Figure 17 shows the small signal pumped spectrum, we focused a 780 nm laser signal into the strip-loaded waveguide cleaved end. The waveguide is pumped with a wavelength of 325 nm and power of 40 mW. Then we collected the output signal which was amplified about 1.03 times.

In an earlier investigation [1], the higher net model gain coefficient was reported to be about 100 cm^{-1}. It is caused by the density of Si nanocrystal and the red shift in the ASE peak wavelength. With a higher density of Si nanocrystal, the quantum confinement effect becomes stronger and leads to higher optical gain of the Si nanocrystal waveguide. In addition, the red shift of the ASE peak wavelength means that there are some luminescent wave losses when propagating in the SiO$_X$ strip-loaded waveguide. On the other hand, the loss of Si nanocrystal was reported to be about 9 cm^{-1},[9] which is caused by the size of Si nanocrystal ranging from 2 to 5 nm in diameter, and also caused by the interface loss at both coupling end-facets. Due to the Si nanocrystal with a larger size in our sample of about 5 nm, the scattering effect in the SiO$_X$

waveguide is inevitably larger than previous reports. Moreover, the slightly large surface roughness of the film grown by PECVD will introduce additional interface scattering loss.

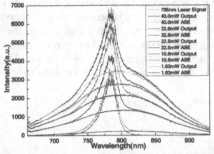

Figure 18. The output spectrum and the ASE spectrum with different pumping power.

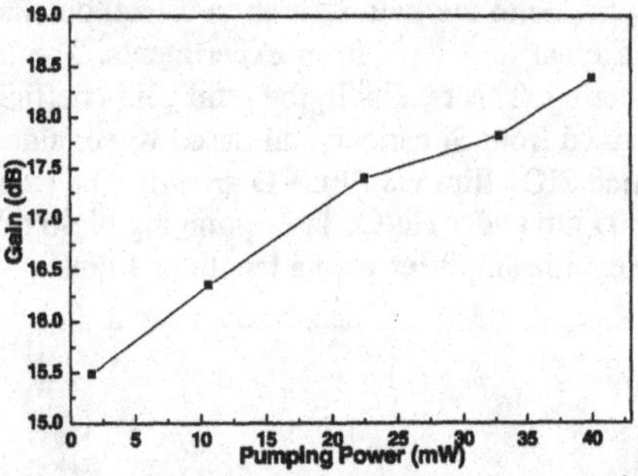

Figure 19. The gain versus the pumping power.

Figure 18 shows the output spectrums with different pumping powers. The laser signal with a wavelength of 785 nm and constant power is the input signal. The wavelength of the pumping source is 325 nm. The ASE spectrum and the output spectrum show that with higher pumping power we can observe higher ASE spectrum peak intensity and output spectrum peak intensity. Figure 19 shows the relationship between the gain and pumping power. The gain coefficient is calculated from the formula as below,

$$Gain = 10 \times \log(\frac{P_{out}}{P_{in}}) - loss$$
$$= 10 \times \log(P_{out}) - 10 \times \log(P_{in}) - loss \qquad (4)$$
$$= 10 \times \log(P_{out}) - const.$$

where P_{out} is output power and P_{in} is input power. The relationship between the pumping power and the optical gain is linear, the optical gain coefficient is higher with higher pumping power. The higher pumping power provides more energy to excite the electrons and stimulates emission from the conduction band to the valence band.

CONCLUSIONS

In conclusion, we fabricated the strip-loaded waveguide with buried Si nanocrystal and determined the optical gain coefficient of the Si nanocrystal to be about 70 cm^{-1}. The optical net model gain and loss coefficient are 65 cm^{-1} and 5 cm^{-1} respectively. The ASE spectrum ranges from 750 to 850 nm with the central wavelength of 805 nm and the 3dB spectral linewidth of 140 nm. The ASE spectrum is red shifted 6 nm to the PL spectrum because of mode guiding. The peak wavelength of the ASE spectrum is blue shifted with longer pumping lengths. Because of the longer pumping lengths, the mode guiding is stronger and the peak wavelength becomes stable. The Si nanocrystal have such a high optical gain coefficient of 70 cm^{-1}, and the Si substrate based characteristic, that these findings could lead the way to a Si based laser. The small-signal power gain of 18.4 dB at the wavelength of 805 nm under He-Cd laser pumping of 40 mW at 325 nm is obtained from the $SiO_2/SiO_x/SiO2$ waveguide amplifier with a length of 1 cm. The relationship between the pumping power and the optical gain is linear, and the optical gain coefficient is higher with higher pumping power.

ACKNOWLEDGMENTS

This work was supported in part by the National Science Council (NSC) of the Republic of China under grants NSC96-2221-E-002-099 and NSC97-ET-7-002-007-ET.

REFERENCES

1. L. Pavesi, L. Dal Negro, C. Mazzoleni, G. Franzò and F. Priolo, "Optical gain in Si nanocrystal," *Nature*, vol. 408, pp. 440-444, (2000).
2. P. Pellegrino, B. Garrido, C. Garcia, J. Arbiol, J. R. Morante, M. Melchiorri, N. Daldosso, L. Pavesi, E. Scheid, and G. Sarrabayrouse, "Low-loss rib waveguides containing Si nanocrystal embedded in SiO_2, " *J. Appl. Phys.*, vol. 97, pp. 074312, (2005).
3. K. Luterová, K. Dohnalová, V. Švrček and I. Pelant, "Optical gain in porous silicon grains embedded in sol-gel derived SiO2 matrix under femtosecond excitation" *Appl. Phys. Lett.*, vol. 84, pp. 3280-3282, (2004).
4. J. H. Shin, J. Lee, H.-S. Han, J.-H. Jhe, J. S. Chang, S.-Y. Seo, H. Lee, and N. Park, "Si nanocluster sensitization of Er-doped silica for optical amplet using top-pumping visible LEDs," *IEEE J. Sel. Top. Quantum Electron*, vol. 12, 783-796, (2006).
5. T. J. Clement, R. G. DeCorby, N. Ponnampalam, T. W. Allen, A. Hryciw and A. Meldrum, "Nanocluster sensitized erbium-doped Si monoxide waveguides," *Opt. Exp.*, vol. 14, pp. 12151-12162, (2006).
6. B. E. Little, Member, IEEE, J. S. Foresi, G. Steinmeyer, E. R. Thoen, S. T. Chu, H. A. Haus, Life Fellow, IEEE, E. P. Ippen, Fellow, IEEE, L. C. Kimerling and W. Greene, "Ultra-compact $Si-SiO_2$ microring resonator optical channel dropping filters," *IEEE Phot. Tech. Lett.*, vol. 10, pp. 549-551, (1998).
7. D. S. Gardner, M. L. Brongersma, "Microring and microdisk optical resonators using Si nanocrystal and erbium prepared using Si technology," *Optical Materias*, vol. 27, pp. 804-811, (2005).
8. K. Jia, W. Wang, Y. Tang, Y. Yang, J. Yang, Member, IEEE, X. Jiang, Member, IEEE, Y. Wu, M. Wang, and Y. Wang, Senior Member, IEEE, "Si-on-insulator-based optical demultiplexer employing turning-mirror-integrated arrayed-waveguide grating," *IEEE Phot. Tech. Lett.*, vol. 17, pp. 378-380, (2005).
9. K. Luterová, M. Cazzanelli, J.-P. Likforman, D. Navarro, J. Valenta, T. Ostatnický, K. Dohnalová, S. Cheylan, P. Gilliot, B. Hönerlage, L. Pavesi, I. Pelant, "Optical gain in nanocrystalline Si: comparison of planar waveguide geometry with a non-waveguiding ensemble of nanocrystal," *Optical Materias*, vol. 27, pp. 750-755, (2005)

Mater. Res. Soc. Symp. Proc. Vol. 1076 © 2008 Materials Research Society   1076-K05-03

Surface Modification of ZrO$_2$ Nanoparticles as Functional Component in Optical Nanocomposite Devices

Ninjbadgar Tsedev[1,2], and Georg Garnweitner[2]

[1]Dept. of Colloid Chemistry, Max Planck Institute of Colloids and Interfaces, Research Campus Golm, Potsdam, 14424, Germany

[2]Institute of Particle Technology, TU Braunschweig, Volkmaroder Str. 5, Braunschweig, 38104, Germany

ABSTRACT

We have recently shown the successful synthesis of zirconia nanoparticles that can be optimized for use in volume phase holography by a post-functionalization surface treatment. Here, we present further investigations on the surface modification treatment with the aim of providing tools to tailor the nanoparticle compatibility to the photocurable organic matrix. Highly crystalline ZrO$_2$ nanoparticles with a mean diameter of 5nm are synthesized in multigram yield through a one-pot solvothermal reaction of zirconium (IV) n-propoxide in benzyl alcohol. It is shown that the yield of the ZrO$_2$ nanoparticles and stability of the nanoparticle dispersions are strongly dependent on the synthesis temperature. Post-synthetic surface modification of ZrO$_2$ nanoparticles using several aliphatic ligands with different surface binding groups such as carboxylate (-COO$^-$), amine (-NH$_2$), phosphate (-PO$_4$) and methoxysilane (-SiOCH$_3$) was performed in order to compare the binding ability of these functional groups to the nanoparticle surface and therefore provide a new rational approach for nanoparticle stabilization with organic ligands.

INTRODUCTION

In the field of optoelectronics and photonics, holography is one of the most dynamic technologies with increasing impact on many applications, including holographic data storage, sensors, security holograms and real-time microscopic imaging [1-3]. Its development hinges on the design and fabrication of volume periodic photosensitive composite materials with high diffraction efficiency, fast response and low optical losses. One promising possibility for fabricating gratings with high refractive index modulation and high stability is the incorporation of inorganic nanoparticles with high refractive index in a photopolymerizable active organic matrix [4-5]. At this point, by taking advantage of their intrinsic high refractive index and transparency, zirconia nanoparticles were successfully explored in an organic-inorganic nanocomposite, leading to effective volume phase holographic gratings after holographic exposure [6]. The crucial parameter for their successful incorporation is their surface modification, which must ensure complete stability and compatibility of the nanoparticles throughout the grating formation process, preventing agglomeration and thus, high scattering in the nanocomposite gratings. Therefore, the future development and application of such nanocomposite holographic devices will crucially depend on the availability of suitable nanoparticles with optimized surface structure [4-8].

Recently we have presented a scalable one-pot solvothermal route to monodisperse, highly crystalline zirconia (ZrO₂) nanoparticles with yields of 20 g per batch [6]. Using fatty acids as surface-modifies, stable dispersions in chloroform were obtained that could be used for the fabrication of nanocomposites in photocurable isooctyl acrylate (IOA)-based organic matrices. Whilst fatty acids with various chain lengths were explored to assess the effect of chain length on the stabilization, the carboxylic acid group as anchor group to the zirconia surface was not varied [6]. Here, we present a study of the effect of different surface binding groups of aliphatic ligands on the particle stabilization both in organic solvents and the IOA host. Carboxylate (-COO⁻), amine (-NH₂), phosphate (-PO₄) and methoxysilane (-SiOCH₃) were used as functional groups, representing the commonly used functional groups for stabilization of metal oxides today. Additionally, we performed investigations on the influence of the synthesis conditions on the particle dispersibility, to gain further insight into the stabilization mechanism.

EXPERIMENTAL DETAILS

For the synthesis, 66.6 g of zirconium (IV) isopropoxide isopropanol complex (Aldrich, 99.9%), or 77.2 ml zirconium (IV) n-propoxide in 1-propanol (70wt %, Aldrich) was added to 500 ml of benzyl alcohol (Aldrich, ≥99%) in a Teflon reaction cup. This was inserted into a 1 l bench top reactor (Parr instruments), which was heated to 210°C under continuous stirring for 3 days. When using zirconium (IV) n-propoxide, the solvothermal reaction was conducted at 220°C to obtain comparable yields. In both cases, the product was isolated as a white powder by centrifugation and washing with ethanol. For the functionalization treatments, the ZrO₂ nanoparticles were not dried after washing but suitable solutions of the ligands (containing 10 mg of ligand/ml CHCl₃) were added immediately after decantation of ethanol, followed by sonication for 20 min.

X-ray diffraction (XRD) measurements were performed on a Bruker D8 diffractometer equipped with a scintillation counter using Cu-K$_α$ radiation with a scanning step size of 0.04°/2θ. The infrared spectra of the samples were recorded using a diffuse reflectance infrared Fourier transform spectroscopy (DRIFTS) Bruker Equinox IFS 55 device, under accumulation of 64 scans. Thermogravimetric analysis (TGA) was performed with a Mettler Toledo TGA/SDTA851 instrument in a flowing oxygen atmosphere employing a heating rate of 10°C min⁻¹.

DISCUSSION

The XRD patterns of ZrO₂ nanoparticles prepared using the two different zirconia precursors are displayed in Figure 1. As reported in our previous work [6], purely cubic zirconia (JCPDC 27-991) is obtained when using zirconium (IV) isopropoxide isopropanol complex as precursor (a). Nonetheless, it is highly attractive to use other precursors such as zirconium (IV) n-propoxide as starting material because of their substantially lower cost. Also when using zirconium (IV) n-propoxide solution in 1-propanol, highly crystalline nanoparticles were obtained as visible in the XRD pattern of the dried nanoparticles in Figure 1 (b). The characteristic ZrO₂ peaks at 30.2°, 35.2°, 50.5° and 60.1° are seen, but due to significant line broadening, it is difficult to assign the peaks specifically to either the cubic or tetragonal phase. Additionally, peaks at 28.1° and at 24.4° and 40.6° are visible that correspond to monoclinic zirconia (JCPDC 37-1484). The particle size was calculated to 2.8 nm for sample (a) using a quantitative line broadening analysis of the (220) diffraction peak, and to 5 nm for (b).

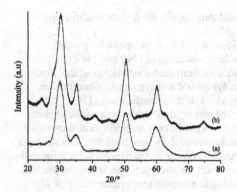

Figure 1. XRD patterns of the ZrO$_2$ nanoparticles prepared from zirconium (IV) isopropoxide isopropanol complex (a) and zirconium (IV) *n*-propoxide (b)

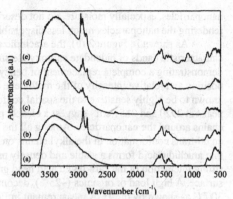

Wavenumber (cm^{-1})

Figure 2. DRIFT spectra of ZrO$_2$ nanoparticles before (a) and after surface modification with arachidic acid (b), dodecylamine (c), OTMS (d) and di(2-ethylhexyl)phosphate (e).

Even without any additional ligand, the as-prepared ZrO$_2$ nanoparticles are slightly dispersible in chloroform, obtaining clear dispersions with solid contents of about 10 mg/ml, owing to organic species bound to the particles that stem from the synthesis, especially the solvent benzyl alcohol. The long term stability of these dispersions however was not good (agglomerates started to form within a day), and the concentration of the nanoparticles in various non-polar organic solvents that need to be used for the preparation of nanocomposites suitable for volume holography is not sufficient to achieve a high diffraction performance. To improve colloidal stability of the ZrO$_2$ nanoparticles in non-polar media, and to gain an insight into the binding capability of various organic functional groups to the zirconia surface, we explored the use of four different aliphatic ligands with various functional groups: arachidic acid, trimethoxy(octyl)silane (OTMS), di(2-ethylhexyl) phosphate and dodecylamine.

In all cases, transparent dispersions with solid nanoparticle contents of about 100 mg/ml were obtained directly after adding the corresponding functionalization solution to the wet nanoparticles after the washing step. Diffuse reflectance infrared Fourier transform spectroscopy (DRIFTS) was performed in order to investigate the binding of the ligands to the nanoparticle surface, using the solid after precipitation with methanol and drying. The obtained spectra of the ZrO$_2$ nanoparticles modified with different ligands are given in Figure 2 and are compared to unmodified ZrO$_2$ nanoparticles. In addition to the characteristic absorption band of Zr-O-Zr vibrations below 600 cm^{-1}, the as-prepared ZrO$_2$ nanoparticles exhibit a series of peaks around 1400-1600 cm^{-1}, corresponding to aromatic benzyl alcohol species adsorbed to the nanoparticles during the synthesis. According to the TGA data presented in Figure 3, the obtained ZrO$_2$ nanoparticles exhibit a weight loss of 8.6%, in a pronounced step between 360 and 460°C, which corresponds to a decomposition/combustion of surface organics. These organic species can make the as-prepared nanoparticles dispersible in certain media, for instance, in chloroform, but also facilitate subsequent post-synthetic surface modification due to easy exchange to surface modifiers. We have recently discovered that other species adsorbed after drying of the nanoparticles, especially moisture, do not desorb and thus impede the surface modification, rendering the nanoparticles much less dispersible [9].

As shown in Figure 2 (b), the arachidic acid-modified ZrO$_2$ nanoparticles exhibit characteristic bands of bidendate carboxylate stretching vibrations at 1548 and 1467 cm^{-1}, demonstrating a complete replacement of benzyl alcohol from the nanoparticle surface. Since position and relative intensity of the methylene antisymmetric and symmetric vibrations are known to be highly sensitive to the spatial conformations of the aliphatic chains [10-11], the peaks at 2920 and 2851 cm^{-1} suggest a closely packed and crystalline-like state of the aliphatic chains around the nanoparticles. Among all the studied ligands, only arachidic acid shows such an all-trans conformation in the alkyl chains on the surface of ZrO$_2$ nanoparticles. This indicates that arachidic acid forms a stable and densely packed monolayer on the nanoparticles through the strong interaction of the carboxylic acid head groups with the zirconium species on the particle surface. A high load of organics (~25%), decomposing in a narrow temperature range of 285-300°C as shown by the TGA measurement in Figure 3 (e), supports this result of a rather densely packed, uniform arachidic acid shell around the nanoparticles.

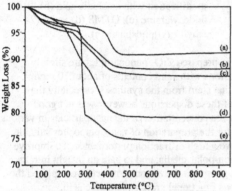

Figure 3. TGA curves of the ZrO$_2$ nanoparticles before (a) and after surface modification with dodecylamine (b), OTMS (c), di(2-ethylhexyl)phosphate (d) and arachidic acid (e).

In the DRIFT spectrum of the OTMS-stabilized nanoparticles, in addition to the stretching vibration of the Si-O bond at 1049 cm^{-1} [12], the peaks from initially adsorbed benzyl alcohol species are still observed at 1400-1600 cm^{-1}, indicating only partial replacement by OTMS. This result is further confirmed by thermogravimetric analysis. Whilst the normal TGA plot (Figure 3 (d)) suggests rather slow decomposition of the organic ligands, in the derivative curve (not shown here), three individual weight loss steps could be distinguished, with a total of 12% organics. A major loss related to decomposition of surface bound ligands occurred in the temperature range of 250-350°C, followed by final loss at 400°C which coincides with that of unmodified ZrO$_2$ nanoparticles, additionally suggesting that benzyl alcohol species are still present after surface modification. From the TGA plot, the ratio of OTMS to benzyl alcohol could be estimated as approx. 2.5. A similar trend was observed for dodecylamine-stabilized ZrO$_2$ nanoparticles (Figure 2 (c)). The N-H deformation vibration of the amine group at around 1600 cm^{-1} is probably overlapping with that of surface organics. The ZrO$_2$ nanoparticles stabilized with di(2-ethylhexyl) phosphate exhibited several DRIFTS absorption peaks at 1022, 1089 and 1191 cm^{-1}, corresponding to P-O and P-O-C stretching vibrations [13]. As can be seen from Figure 3 (d), the thermogram of ZrO$_2$ nanoparticles stabilized with di(2-ethylhexyl) phosphate shows a gradual decrease in weight up to 500°C, with a final loss of 17.6%, which corresponds to decomposition of surface bound ligands, proving the binding of di(2-ethylhexyl) phosphonate to the ZrO$_2$ surface.

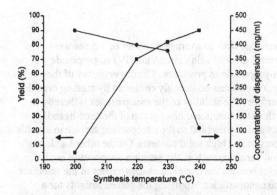

Figure 4. Dependence of reaction yield and dispersibility on the synthesis temperature

Interestingly, the yield of nanoparticles from the solvothermal synthesis as well as their dispersibility in particular media are strongly dependent on the reaction conditions. Here, we discuss the dependence of reaction yield and dispersibility on the synthesis temperature. To study the dispersibility of the nanoparticles after post-functionalization with arachidic acid, the maximum concentration of the nanoparticles in homogeneous dispersion were determined by preparing concentrated suspensions in the functionalization solution. After sonication for 20 min, the turbid liquids were centrifuged in order to remove all agglomerated nanoparticles, resulting in transparent dispersions in all cases. The solid content of these dispersions was determined gravimetrically by heating at 400°C. As shown in Figure 4, as the reaction temperature rises from 200 to 240°C, the obtained yield of nanoparticles increases drastically. At the same time, the maximum solid content of the dispersions decreases significantly above 230°C. The relatively low yield of ZrO_2 nanoparticles at a reaction temperature of 200°C may however be partly due not to incomplete reaction, but to a good stabilization of the nanoparticles also in benzyl alcohol, therefore making them impossible to separate from the solvent by centrifugation (as was performed here). This is supported by the increasing stability of the nanoparticles after post-functionalization, which possibly points to a high stability also before the functionalization. At higher synthesis temperatures, we infer that the number of surface binding sites of the zirconia nanoparticles is reduced, possibly due to enhanced ripening and elimination of surface defects of the nanocrystals.

In order to assess the capability of the ligands for stabilization in photocurable isooctyl acrylate (IOA)-based organic matrices for the preparation of volume holographic devices, we added dispersions of the stabilized nanoparticles to the IOA monomer, followed by evaporation of the volatile organic solvent to obtain nanoparticle-IOA composites with a ZrO_2 content up to 15 wt%. All dispersions can initially be mixed with IOA to obtain transparent liquids. The dispersions with dodecylamine and di(2-ethylhexyl) phosphate-stabilized nanoparticles however became turbid within 30 minutes, showing that these only weakly attached ligands are desorbed quickly due to destabilization of the bond between their functional head groups and the zirconia surface when a third component is introduced to the system, consequently leading to agglomeration of the ZrO_2 nanoparticles. The arachidic acid and OTMS-stabilized nanoparticles remained stable and non-agglomerated in the IOA matrix even after complete evaporation of the solvent, demonstrating their strong interaction with zirconia through multidendate coordination, as already confirmed with DRIFTS and TGA. Clearly, the arachidic acid and OTMS-stabilized ZrO_2 nanoparticles exhibit better compatibility with IOA owing to their stronger binding affinity for zirconia.

CONCLUSIONS

Highly crystalline zirconia nanoparticles with an average diameter of 3-5 nm were synthesized via a non-aqueous route in benzyl alcohol using zirconium (IV) isopropoxide isopropanol complex or zirconium (IV) n-propoxide as precursor. The dispersibility of the resulting ZrO_2 nanoparticles in non-polar solvents can be strongly enhanced by binding organic stabilizers (ligands) to the nanoparticle surface. The stability of the nanoparticles is thereby strongly dependent on the binding strength of the functional head group of the used ligand. Chelating ligands such as carboxylic acids and silanes bind to the nanoparticle surface in a stable fashion and therefore result in stable dispersions at high solid contents. On the other hand, our results highlight the importance of reaction variables such as the reaction temperature on the performance of resultant ZrO_2 nanoparticles besides the chemical characteristics of the stabilizer for achieving optimum stabilization of the nanoparticles. Applying the gained insights for a tuning of the synthesis and post-functionalization treatments, highly stable nanocomposites of the nanoparticles can also be achieved in the photocurable organic monomer IOA, providing means for the optimization of the nanoparticle components for efficient holographic gratings.

ACKNOWLEDGMENTS

We thank Prof. Dr. H. Menzel, Institute of Technical Chemistry, TU Braunschweig, for use of the DRIFTS instrument. N.T. thanks the Alexander von Humboldt Foundation for a research fellowship.

REFERENCES

1. A.S. Matharu, S. Jeeva and P.S. Ramanujam, *Chem. Soc. Rev.* **36**, 1868 (2007).
2. C. Gu, Y. Xu, Y. Liu, J.J. Pan, F. Zhou, H. He, *Optical Mater.* **23**, 219 (2003).
3. V.A. Barachevskii, *High Energy Chem.* **40**, 165 (2006).
4. C. Sánchez, M.J. Escuti, C. van Heesch, C.W.M. Bastiaansen, D.J. Broer, J. Loos and R. Nussbaumer, *Adv. Funct. Mater.* **15**, 1623 (2005).
5. O.V. Sakhno, T.N. Smirnova, L.M. Goldenberg and J.Stumpe, *Mater. Sci. and Eng. C*, **28**, 28 (2008).
6. G. Garnweitner, L.M. Goldenberg, O.V. Sakhno, M. Antonietti, M. Niederberger and J.Stumpe, *Small*, **3**,1626(2007).
7. K. Saravanamuttu, C.F. Blanford, D.N. Sharp, E.R. Dedman, A.J. Turberfield and R.G. Denning, *Chem. Mater.* **15**, 2301 (2003).
8. N. Suzuki, Y. Tomita and T. Kojima, *Appl. Phys. Lett.* **81**, 4121 (2002).
9. S. Zhou, G. Garnweitner, M. Niederberger, M. Antonietti, *Langmuir* **23**, 9178 (2007).
10. C.K. Yee, A. Ulman, J.D. Ruiz, A. Parikh, H. White and M. Rafailovich, *Langmuir*, **19**, 9450 (2003).
11. M.J. Hostetler, J.J. Stokes and R.W. Murray, *Langmuir* **12**, 3604 (1996).
12. R. De Palma, S. Peeters, M.J. van Bael, H. van den Rul, K. Bonröy, W. Laureyn, J. Mullens, G. Borghs and G. Maes, *Chem. Mater.* **19**, 1821 (2007).
13. G. Guerrero, P.H. Mutin and A. Vioux, *Chem. Mater.* **13**, 4367 (2001).

Mater. Res. Soc. Symp. Proc. Vol. 1076 © 2008 Materials Research Society   1076-K06-06

Temperature Dependent Fluorescence of Nanocrystalline Ce-doped Garnets for Use as Thermographic Phosphors

Rachael Hansel[1], Steve Allison[2], and Greg Walker[3]

[1]Interdisciplinary Graduate Program in Materials Science, Vanderbilt University, Nashville, TN, 37212
[2]Oak Ridge National Laboratory, Oak Ridge, TN, 37831-6054
[3]Department of Mechanical Engineering, Vanderbilt University, Nashville, TN, 37212

ABSTRACT

Four samples of $(Y_{1-x}Ce_x)_3(Al_{1-y}Ga_y)_5O_{12}$ (where x=0.01, 0.02 and y=0, 0.5) were synthesized via the simple, efficient combustion synthesis method in order to determine the effect of substituting Ga^{3+} for Al^{3+} on the temperature-dependent fluorescent lifetime. X-ray diffraction shows that the Ga-doped samples have longer lattice constants and transmission electron microscopy data show that each sample consists of nanocrystallites which have agglomerated in micron-sized particles. Photoluminescence data reveal that the addition of gallium into the YAG:Ce matrix induces a red shift in the absorption spectra and a blue-shift in the emission spectra. The laser-induced fluorescent lifetime was determined as a function of temperature over the range of 0-125°C using two different emission filters. Increasing the amount of dopant ultimately results in a decrease of the fluorescent lifetime. The quenching temperatures for the Ga-doped samples were lower than the samples without gallium. The results of this work show that combustion synthesis is viable method for making highly luminescent, nanocrystalline TGPs. In addition, these results show that the quenching temperature of YAG:Ce can be altered by substituting ions which alter the location of the charge transfer state and by changing the morphology of the sample.

INTRODUCTION

Cerium-doped yttrium aluminum garnet ($Y_3Al_5O_{12}$:Ce, YAG:Ce) is used in several applications such as solid state lighting, displays, and scintillators [18], [3], [16]. The Ce^{3+} ion is responsible for nanosecond decay time and an intense visible emission. Cerium-doped garnets are also being considered for use as a thermographic phosphor. Thermographic phosphors are a special class of materials commonly used as non-contact thermometers because the fluorescent decay lifetime is temperature dependent [3]. The luminescent properties of doped nanocrsytals are attracting great interest now because the fluorescent characteristics of nanocrystalline phosphor materials change with particle size. For example, the decay time of bulk YAG:Ce particles is temperature dependent between 150-300°C. However, Allison et al. [4] showed that nanocrystalline YAG:Ce exhibits temperature dependency between 7-77°C. The shift in temperature dependency is observed as a function of size and has also been observed in Y_2O_3:Eu nanocrystals. The size effect is likely due to increasing lattice distortions in smaller particles [12]. Similarly, the quenching temperature of bulk yttrium aluminum garnet, where Ga^{3+} is substituted for Al^{3+} and Ce^{3+} is substituted for Y^{3+} (YAGG:Ce), is approximately 100°C and shows microsecond decay times [2]. However, there are no studies which determine the temperature dependency of nanoparticles of YAGG:Ce.

In this work, we have synthesized nanocrystalline particles of YAGG:Ce in order to determine how gallium affects the temperature-dependent fluorescent properties of YAG:Ce. We have determined the fluorescent lifetime as function of temperature and analyzed the results in relation to crystal structure and the location of the charge transfer state. The results of this study show how nanocrystalline features can be exploited to design other phosphors for use as TGPs. To the best of our knowledge, there has been no study that compares the performance of YAG:Ce and YAGG:Ce in relation to crystal structure and particle size for use as a thermographic phosphor.

EXPERIMENTAL

Four samples of $(Y_{1-x}Ce_x)_3(Al_{1-y}Ga_y)_5O_{12}$ where x=0.01, 0.02 and y=0, 0.50, were made by combustion synthesis in which aqueous solution of stoichiometric molar ratios of $Y(NO_3)_3$ (anhy), $Al(NO_3)_3$ *$9H_2O$, $Ce(NO_3)_3$ *$6H_2O$, $Ga(NO_3)_3$ *$9H_2O$, and $CO(NH_2)_2$ (urea). Each sample was placed in a muffle furnace at 500°C to evaporate the water, after which the auto-combustion process began with the evolution of a white gas. Immediately following the auto-combustion, a voluminous, porous yellow powder formed. All samples were calcined for 5 hours at 1000°C, which is well above the reported crystallization temperature of 900°C [18], [11]. X-ray powder diffraction (XRD) spectra were measured at room temperature with a Scintag XRD and the average crystallite size was determined by using Scherrer's equation. Transmission electron microscopy (TEM) images were collected on a Hitachi 3300 with 100kV beam. Room temperature photoluminescence measurements were taken using a QuantaMaster-14/2005 with a 150W Xenon lamp excitation source.

The temperature dependence was determined by using the configuration shown in figure 1.

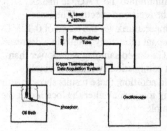

Figure 1. Fluorescent Temperature Dependent Experimental Setup

The excitation source was a nitrogen laser (Laser Science Corporation, model VSL-337ND) with λ_{ex}=337nm and an excitation band width of 0.1nm. The pulse width was 4ns at a characteristic energy of 300μJ. The excitation pulse was conveyed to the sample via a 1mm optical fiber. An identical fiber collected and transmitted the emitted signal to a photomultiplier tube which served as the detector. Each phosphor sample was placed in the bottom of the plastic capsule which covered the excitation and detector fiber. The capsule/fiber was placed in an oil bath and slowly heated at a rate of 1°C/min and the temperature of the oil bath was monitored by a k-type bare wire thermocouple. Bandpass filters centered at 540nm and 700nm were used to collect the emitted signal. A waveform processing oscilloscope with 350Hz bandwidth displayed, digitized, and stored the data.

RESULTS

Structure

The X-ray diffraction (XRD) spectrum of all calcined samples is shown in figure 2. All peaks approximately match the cubic $Y_3Al_5O_{12}$ phase (JCPDS file 33-40 [1]). The presence of gallium in the YAG matrix shifts the peak positions to lower diffraction angles due in part to the fact that the radii of Ga^{3+} and Al^{3+} ions are 0.062nm and 0.053nm, respectively [15]. Since Ga^{3+} is larger than the host site, the inter-planar spacing will decrease and the lattice constant will increase due to the increased repulsion between adjacent atoms (Table 1).

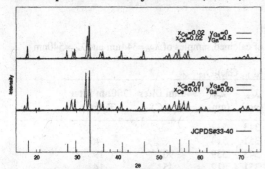

Figure 2. XRD spectra of calcined samples

Transmission electron microscopy images of the calcined 1%Ce and 0%Ga (x=0.01and y=0.5) are shown in figure 3.

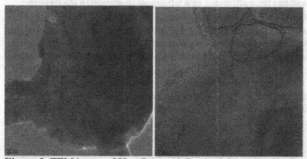

Figure 3. TEM image of $Y_{2.97}Ce_{0.03}Al_5O_{12}$ (x=0.01 and y=0)

The material is composed of micron-sized agglomerates composed of ~30nm crystallites. In addition, the crystallites are embedded in amorphous material. The source of the amorphous material is most likely due to inhomogeneous flame-temperature distribution during the synthesis. Future work will use energy dispersive spectroscopy techniques to identify and quantify the amorphous components of each sample.

Photoluminescence

Photoluminescence spectra are shown in figure 4 and corresponding data are given in Table 1.

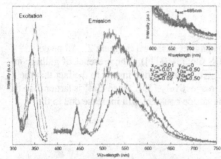

Figure 4. Excitation and emission spectra of calcined samples of λ_{exc}=344nm and λ_{em}=540nm

Table 1. Experimental Results $(Y_{1-x}Ce_x)_3(Al_{1-y}Ga_y)_5O_{12}$

x (%Ce)	y (%Ga)	a (nm)	Average crystallite size (nm)	λ_{exc} (nm)	λ_{em} (nm)	540nm filter T_q (°C)	700nm filter T_q (°C)
0.01	0	1.209	37.11	343	537	-	-
0.01	0.5	1.222	27.07	351	514	-	~45
0.02	0	1.210	32.43	343	539	90	19
0.02	0.5	1.227	27.28	351	517	45	14

The addition of gallium into YAG:Ce red-shifts the absorption peak approximately 8nm while the emission peak is blue-shifted approximately15nm. The blue-shift of the emission wavelength correlates well with other studies and has been attributed to a tetragonal distortion of the oxygen atoms surrounding the Ce^{3+} ion brought about by the substitution of Al^{3+} for Ga^{3+} [17] [6]. The inset of figure 4 shows a portion of the emission spectra using λ_{exc}=485nm to excite the sample. Using this excitation wavelength, a small, low-intensity emission was observed at 700nm. There are no known emission peaks for Ce^{3+} in the red region. Trivalent chromium is known to absorb in the blue region and emit around 700nm due to the $^4T_2 \rightarrow {}^4A_2$ and $^2E \rightarrow {}^4A_2$ transitions [14]. It is highly likely that this anomalous emission is due to Cr^{3+} contamination because the doublet shape, multi-exponential decay curves, and emission lifetime are very similar to known emission spectra of YAG:Cr [14]. As shown below, this unexpected emission was very useful in demonstrating the effect of Ga-substitution on the quenching temperature. In the future, atomic absorption spectroscopy will be used to verify Cr^{3+} contamination.

Fluorescence Temperature Dependency
Two bandpass filters centered at 540nm and 700nm were used to determine the temperature dependency. These emission filters were chosen because of emission signals given in the PL data. A 700nm filter was chosen because spectral data showed a low-intensity, broad-band emission in the red region. Figure 5 shows the fluorescent decay lifetime as a function of temperature. Quenching temperatures and lifetime ranges are shown in table 1.

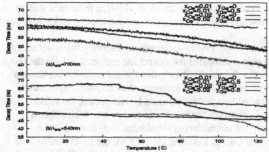

Figure 5. Decay Time vs. Temperature of $(Y_{1-x}Ce_x)_3(Al_{1-y}Ga_y)_5O_{12}$ using an (a)700nm and (b)540nm emission filter.

The quenching temperature, T_q, is the temperature at which fluorescence begins to decrease due thermal effects within the lattice. In addition, the lifetimes of the nanocrystals were approximately 60ns which correlates well with observed lifetimes in both micron and nanosized Ce-doped phosphors [4] [5]. It was interesting to observe that lower quenching temperatures were observed when using a red emission filter.

DISCUSSION

From the data presented in figure 5, it is clear that the quenching temperatures for the samples with Ga^{3+} are at lower temperatures compared to the samples without Ga^{3+}. The reasons for lower T_q will be discussed in relation to the particle morphology and to changes induced in the host lattice as a result of Ga-doping.

Figure 5a and 5b show that the samples with no Ga-doping exhibit a very small variation in lifetime. It is possible that the crystallite size is too large to observe strong temperature dependence in the range of 0-120°C (see table 1). In this case, the temperature dependent behavior of this material is similar to bulk YAG:Ce where the quenching temperature is 150°C. In contrast, the average crystallite size of the Ga-doped samples is noticeably smaller than the sample with no Ga^{3+}. This indicates that the addition of dopants into the lattice introduces defect centers into the crystal which prohibit crystal growth. Temperature-dependence measurements of the luminescence of YAG:Ce have shown that energy trapped at defect centers can be released thermally which leads to competition for radiative emission between activator and defect centers [13]. The number of pathways available for non-radiative transitions increases when defects (in the form of dopants) are introduced into the crystal. Thus, Ga-doping in nanocrystallites introduces defects into the lattice which reduce the crystallite size and ultimately serve as non-radiative pathways which become energetically favorable at lower temperatures which result in lower quenching temperatures.

In addition to the size effect, the presence of Ga^{3+} also influences the electronic states of the surrounding atoms, specifically the charge transfer state (CTS) of the Ce-O bond. Lower T_q values indicate that the energetic location of the CTS is at a lower energy when Ga^{3+} is present. This can be explained by considering the lattice structure of the YAG matrix in relation to the covalency of Ga-O bond versus the Al-O bond. Nakatsuka et al. [10] reported that Ga^{3+} ions preferentially occupy the tetrahedral site. The strong covalency of the Ga-O bond introduces shielding effects by electrons which reduce cation-cation repulsive forces and stabilize the crystal matrix. As a result of the shielding effects, the repulsion between the octahedrons on either side of the tetrahedron is reduced and the octahedrons will move closer together. Because the octahedrons are closer, the electronic interaction between the Ce^{3+} and the octahedral oxygen atom will decrease and the Ce-O bond will increase in covalency [7]. Consequently, the charge-transfer transitions between these ions shift to lower energy. When the CTS is at a lower energy, lower temperatures are required to make this state energetically favorable.

Finally, the quenching temperatures are considerably lower when observing the lower energy transition (collecting emission at 700nm). Table 1 compares the quenching temperatures for the Ga-doped samples at the two emission wavelengths. It is highly likely that the red emission occurs as a result of Cr^{3+} contamination. The decay curve for the 700 nm emission showed multi-exponential decay and showed low temperature T_q values. Although this emission is unexpected, the effect of Ga-doping lowering the CTS (described above) is observed for this lower energy transition. This is interesting because Cr^{3+} ions will substitute in the octahedral or tetrahedral sites instead of the dodecahedral sites. It is not clear at this point if Ga^{3+} is lowering the CTS for the Cr-O transition or an energy transfer is occurring between the two dopant ions, however, it is clear that Ga^{3+} lowers the quenching temperature.

CONCLUSIONS

The fluorescent lifetime of nanocrystallite $(Y_{1-x}Ce_x)_3(Al_{1-y}Ga_y)_5O_{12}$ has been investigated as a function of temperature between 0-120°C. The results of this work show that nanocrystals of YAGG:Ce synthesized via the simple combustion synthesis technique can be used as thermographic phosphors for low temperature non-contact remote-sensing thermometry. The quenching temperatures of Ga-doped samples were lower than the samples with no gallium because the CTS is lower in the Ga doped samples. It has been demonstrated that T_q can be altered by doping multiple sites in the garnet lattice and by altering the morphology (particle size) of the phosphor. These results may be useful in the design and implementation of thermographic phosphors and also phosphors used in solid-state lighting, display, or light-emitting diodes.

REFERENCES
1. [1] JCPDS Powder Diffraction Standards Data Base, 1990.
2. S.W. Allison, M. R. Cates, and D. L. Beshears., May 2000.
3. S.W. Allison and G.T. Gillies. Rev. Sci. Inst, 68(7), 2615 (1997).
4. S.W. Allison, G.T. Gillies, A. J. Rondinone, and M. R. Cates. Nanotechnology. 14, 859 (2003).
5. R. Asakura, T. Isobe, K. Kurokawa, T. Takagi, H. Aizawa, and M. Ohkubo. J.of Lum. 126, 416 (2007).
6. G. Blasse and A. Bril. J. of Chem. Phys. 47(12), 5139 (1967).
7. G. Blasse and B.C. Grabmaier. Luminescent Materials, (Springer-Verlag, Berlin1994).
8. A. Katelnikovas, P. Vitta, P. Pobedinskas, G. Tamulaitis, A. Zukauskas, J. Jorgensen, and A. Kareiva. J.of Crys.Gro., 304(2), 361 (2007).
9. J. Lin and Q. Su. J.of Mater. Chem. 5(8), 1151 (1995).
10. A. Nakatsuka, A. Yoshiasa, and T. Yamanaka. Acta Crys.B. 55, 266 (1999).
11. X. Pan, M.Wu, and Q. Su. Mat. Sci.and Eng. B, 106, 251 (2004).
12. H. Peng, H. Song, B. Chen, J. Wang, S. Lu, and X. Kong. J. of Chem. Phys. 118(7), 3277 (2003).
13. D.J. Robbins, B. Cockayne, B. Lent, C.N. Duckworth, and J.L. Glasper. Phys. Rev. B, 19(2), 1254 (1979).
14. Y. R. Shen and K.L. Bray. Phys. Rev. B, 56(17) 10882 (1997).
15. P.Y. Jia. Thin Solid Films. 483, 122 (2005).
16. J. Tous, K. Blazek, L. Pina, and B. Sopko. Rad. Meas., 42, 925 (2007).
17. J.L. Wu, G. Gundiah, and A. Cheetham. Chem. Phys. Let. 441, 250 (2007).
18. G. Xia, S. Zhou, J. Zhang, and J. Xu. J. Crys. Gro. 279, 2005.

Mater. Res. Soc. Symp. Proc. Vol. 1076 © 2008 Materials Research Society   1076-K09-02

Non-selective optical wavelength-division multiplexing devices based on a-SiC:H multilayer heterostuctures

Manuela Vieira[1,2], Miguel Fernandes[1], Paula Louro[1,2], Manuel Augusto Vieira[1,3], Manuel Barata[1,2], and Alessandro Fantoni[1]

[1]Electronics Telecommunication and Computer Dept., ISEL, Rua Conselheiro Emídio Navarro, Lisbon, 1959-007, Portugal
[2]CTS, UNINOVA, Monte da Caparica, Caparica, 2829-516, Portugal
[3]Traffic Dept., CML, Lisbon, Portugal

ABSTRACT

In this paper we present results on the optimization of multilayered a-SiC:H heterostructures for wavelength-division (de) multiplexing applications. The non selective WDM device is a double heterostructure in a glass/ITO/a-SiC:H (p-i-n) /a-SiC:H(-p) /a-Si:H(-i')/a-SiC:H (-n')/ITO configuration. The single or the multiple modulated wavelength channels are passed through the device, and absorbed accordingly to its wavelength, giving rise to a time dependent wavelength electrical field modulation across it. The effect of single or multiple input signals is converted to an electrical signal to regain the information (wavelength, intensity and frequency) of the incoming photogenerated carriers. Here, the (de) multiplexing of the channels is accomplished electronically, not optically. This approach offers advantages in terms of cost since several channels share the same optical components; and the electrical components are typically less expensive than the optical ones. An electrical model gives insight into the device operation.

INTRODUCTION

Until the late 1980s, optical fiber communications was mainly confined to transmitting a single optical channel. Because fiber attenuation was involved this channel required periodic regeneration, which included detection, electronic processing, and optical retransmission. Such regeneration causes a high-speed optoelectronic bottleneck and can handle only a single wavelength. Recently plastic optical fibers (POF) [1] are drawing the attention of the industry because they are produced at very low cost and easily installed by non-specially trained people both in home networking, industrial control networks and automotive applications [2]. To improve the transmission data rate, Wavelength Division Multiplexing (WDM) can be employed. WDM enables the use of a significant portion of the available fiber bandwidth by allowing many independent signals to be transmitted simultaneously on one fiber, with each signal located at a different wavelength. Routing and detection of these signals can be accomplished independently, with the wavelength determining the communication path by acting as the signature of the origin, destination or routing. Components are therefore required to be wavelength selective, allowing for the transmission, recovery, or routing of specific wavelengths. Although they are well known for infrared telecommunication, they must be completely renewed for the transmission with the POF fibers. So, the conception of new devices for signal (de)multiplexing in the visible spectrum is a demand in this field [3, 4, 5]. This paper presents preliminary results on the applicability of multilayered a-SiC:H heterostructures as electrically programmable optical filters for WDM tunable devices.

EXPERIMENTAL DETAILS

The device is a double heterostructure in a glass/ITO/a-SiC:H (p-i-n)/ a-SiC:H(-p)/a-Si:H(-i')/a-SiC:H(-n)/ITO configuration produced by Plasma Enhanced Chemical Vapor Deposition (figure1). The thickness and the absorption coefficient of the front p-i-n photodiode are optimized for blue collection and red transmittance and the thickness of the back one adjusted to achieve high collection in the red spectral range. So, both front and back diodes work as optical filters confining the blue (λ_B=450 nm) and the red (λ_R=650 nm) photogenerated carriers, respectively, at the front and back diodes [6].

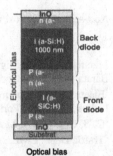

Figure 1. WDM device configuration.

Monochromatic beams together or one single polychromatic (mixture of different wavelength) beam impinge on the device and are absorbed, depending on their wavelength, giving rise to a time and wavelength dependent electrical field modulation across it [7].

The combined effect of the input signal is converted to an electrical signal, via the device, keeping the input information (wavelength, intensity and modulation frequency). By reading out, under different electrical bias conditions, the photocurrent generated by the incoming photons, the input information is electrically multiplexed or demultiplexed and can be transmitted again. In the multiplexing mode the device faces the modulated light incoming together from the fibers (monochromatic input channels) and the electronic signal is read out, under reverse bias. In the multiplexing mode a polychromatic modulated light beam is projected onto the device and the readout is performed by shifting between forward and reverse bias.

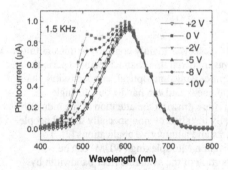

Figure 2. Spectral photocurrent under different applied voltages and modulated frequency of 1.5 kHz.

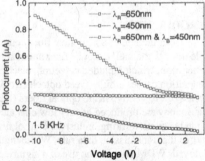

Figure 3. The ac IV characteristics under λ_R=650 nm, λ_B=450 nm, λ_R=650 nm & Bλ_B=450 nm modulated light at 1.5 kHz.

The devices were characterized through spectral response (400-800nm) and photocurrent-voltage (-10V <V <+2V) measurements. In figure 2 the spectral photocurrent measured at 1.5 kHz is displayed at different applied voltages. In figure 3 the *ac* current-voltage characteristics under illumination are shown. In this measurement two modulated (1.5 kHz) monochromatic lights; red (λ_R=650 nm) and blue (λ_B=450 nm) and a polychromatic one (λ_R=650 nm & λ_B=450 nm) where impinging the device one after the other and the photocurrent was measured as a function of the voltage.

Results show that, as the applied voltage changes from forward to reverse the blue/green spectral collection is enlarged while the red one remains constant (figure 2). The collection efficiency (figure 3) under red modulated light is independent on the applied voltage while under blue or combined red and blue irradiation, it quickly increases under reverse bias. It is interesting to notice that, under forward bias, the red and the red & blue signals are almost the same. This behavior illustrates the blindness of the device, under forward bias, to the blue component of the multiplexed signal.

RESULTS AND DISCUSSION

Wavelength division (de)multiplexing

The effect of the applied voltage (-5V<V<+2V) on the output transient multiplexed signal was analyzed (figure 4). The two input wavelength channels are superimposed in the top of the figures to guide the eyes. The frequency of one was 1.5 kHz and the other always half of this value. In figure 4a the signals are due to the combined effect of two beams either with the same wavelength, R&R (λ_R=650 nm); B&B (λ_B=450 nm), or distinct wavelengths, R&B (λ_R=650 nm & λ_B=450 nm). The arrows are indicative of the external voltages: -5 V (solid), 0V (dash), +2V (dot). In figure 4b the multiplexed R&B signal is shown for electrical bias between +2 V and -5 V.

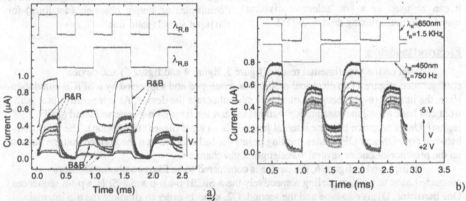

Figure 4. Multiplexed signals at different applied voltages (-5V<V<+2V) and input wavelengths: a) R&R ($\lambda_{R,R}$=650,650 nm); B&B ($\lambda_{B,B}$= 450,450 nm), R&B($\lambda_{R,B}$=650,450 nm). b) R&B ($\lambda_{R,B}$=650,450 nm). The frequency of one of the input signal was 1.5 kHz and the other was half.

Both figures show that the multiplexed signal depends on the applied voltage and on the wavelength of the input channels. If both input channels have the same wavelength (figure 4a) two opposite behaviors are observed. Under red irradiation (R&R) the signal is high and does not depend on the applied bias, under blue irradiation (B&B) the signal increases with the negative bias and in forward bias is irrelevant. A good linearity in the signal amplitude is observed in both

189

situations, showing that the joint effect of the input channels is cumulative. If both channels have different wavelength (R&B, figure 4a and figure 4b), under reverse bias, there are always four levels. The highest level when both red and blue channels are ON and the lowest if both are OFF. In between, the red level (blue channel OFF) is higher than the blue level (red channel OFF). The step among them depends on the applied voltage and increases with the increase of the reverse bias. Under forward bias the blue signal goes down to zero (figure 2, figure 3 and figure 4) and the red one remains constant. The device is blind to the blue irradiation and, thus, the red signal is recovered. Results have shown that different wavelengths which are jointly transmitted must be separated to retain all the information.

The generated photocurrent was measured at two applied voltages to read out the combined spectra. The result is plotted in figure 5, in short circuit (open symbol) and under reverse bias (line).Results show that under short circuit the blue component of the combined spectra falls into the dark level, tuning the red input channel. Thus, by switching between short circuit and reverse bias the red and the blue channels were recovered.

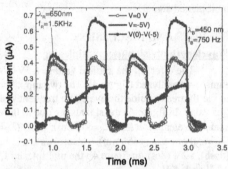

The device acts as a charge integrator, keeping the memory of the input channel. So, it can be used as a non selective division wavelength multiplexing device, WDM.

Figure 5. Blue and red wavelength division demultiplexing output channels (dot lines) for the input signal (solid line).

Electrical modeling

Based on the experimental results (figure 3, figure 4 and figure 5) and device configuration (figure 1) an electrical model was developed and supported by a SPICE simulation. Here, the internal (n-p) junction controls the current across the device. At low positive bias both front and back p-i-n junctions are forward-biased and the internal n-p reverse biased (OFF region). Under negative bias the internal junction is forward-biased (ON region). The transition between both ON and OFF states (turning point) depends not only on the applied voltage but also on the presence of one or several wavelength input channels.

As displayed in figure 6, the device is considered to be like two phototransistors connected back to back, modeling respectively the a-SiC:H p-i-n-p and a-Si:H n-p-i-n sequences. One transistor, Q1, is *pnp* type and the second, Q2, *npn*. In order to simulate the n-p internal junction, the *pnp* shares its collector with the base of the *npn* and its base with the *npn* collector. Capacitors, C1 and C2, are used to simulate the capacitance transient due to the minority carrier trapped in both p-i-n junctions. A voltage source has been applied through a resistor, giving rise to a current I (R3). Two *ac* current sources, I1 and I2, are used to simulate the input blue and red channels photocurrents. The frequencies are the same as the ones used in the experimental work (figure 3a). In figure 6b the input channels, I(I1), I(I2), the simulated multiplexed signal, I(R3) and the current across the capacitors, I(IC1), I(IC2) are shown. A good agreement between experimental and simulated results is achieved (figure 4).

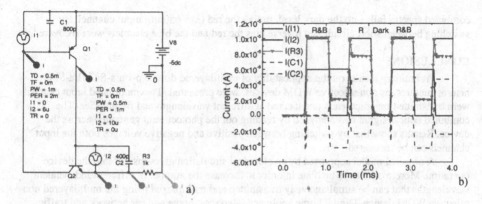

Figure 6. a) Equivalent electrical circuit of the pin-pin photodiode. b) Signals obtained using SPICE simulation when the red (I2) and the blue (I1) modulated lights are impinging the device.

Results show that if the device is biased negatively (-5V) Q1 and Q2 are in their reverse active regions. The p-n internal junction is forward-biased and the external voltage drops mainly across both front and back reverse-biased junctions, mainly at the front one due to its higher resistivity. So, the external current, I, depends not only on the balance between both blue and red photocurrents (I1, I2) [8] but also on the end of each half-cycle of each modulated current. Here, the movement of charge carriers with an increase/decrease in the irradiation, results in a charging or a displacement current similar to the current (i=CdV/dt) that charges the capacitors C1 and C2 in opposite ways.

In the beginning of the cycle (R&B), I1 flows across Q1 collector towards the base of Q2 and together with the photogenerated carriers by the red light recombine or are collected (R&B level). Under blue irradiation, I(I2)=0, only the carriers generated by the blue photons are injected into the base of C1 (B level). C1 charges positively and C2 negatively as a reaction to the decrease in the red irradiation. The opposite occurs under red irradiation where only the red photogenerated carriers contribute to the external current (R level). Here, the capacitors recharge in an opposite way. When both red and blue lights are simultaneously OFF, the current is limited by the leakage current of both active junctions (dark level). So, once triggered, the device continues to conduct until the current through it drops below a certain threshold value, such as at the end of a half-cycle, keeping the information of the wavelength (R&B, R, B, Dark) and frequency (f1, f2) of the impinging light. When a positive voltage is applied to turn the device from the OFF to the ON regions, the junction capacitance across the internal n-p junction is charged. The charging current flows through the emitter of the two transistors. The device behaves essentially as a *npn* phototransistor with the *pnp* transistor acting like a emitter-follower with a very small gain. So, under lower positive voltages the only carriers collected come from the red channel enabling the demultiplexing of the previous multiplexed signal (Fig.5).

Comparing both the experimental and the simulated results it is observed that, under negative applied voltages, the multiplexed signal keeps the memory of the single input channels. It presents four distinct levels, ascribed to the simultaneous presence of both (λ_1; λ_2); one (λ_1;0) / (0;λ_2;) or none (0;0) optical carriers. Under positive bias the blue (λ_1= 450nm) component of the combined spectra falls into the dark level, tuning the red (λ_2= 650nm) input channel. By switching between short circuit and reverse bias the red and the blue channels were recovered.

CONCLUSIONS

Preliminary results on the applicability of multilayered double p-i-n a-SiC:H/a-Si:H heterostructures, as non selective WDM devices, were presented. Two modulated input channels were transmitted together, each one located at different wavelength and frequencies. The combined optical signal was analyzed by reading out the photocurrent generated across the device. Results show that by switching between positive and negative voltages both the input channels can be recovered.

A physical model supported by an electrical simulation gives insight into the device operation. More work has to be done in order to increase the number of different independent wavelengths that can be simultaneously transmitted and multiplexed using the multilayered non selective WDM device. Digital home appliance interfaces, home and car network and traffic control applications are foreseen due to the low cost associated to the amorphous a-SiC:H and POF technologies.

ACKNOWLEDGMENTS

This work has been financially supported by POCTI/FIS/58746/2004 and Fundação Calouste Gulbenkian. The authors tank G. Lavareda and N. de Carvalho by the device deposition.

REFERENCES

1. M. G. Kuzyk, Polymer Fiber Optics, Materials Physics and Applications, Taylor and Francis Group, LLC; (2007).
2. M. Kagami. "Optical Technologies for Car Applications Innovation of the optical waveguide device fabrication". Optical communications - perspectives on next generation technologies, October 23-25, 2007 in Tokyo, Japan.
3. S. Randel, A.M.J. Koonen, S.C.J. Lee, F. Breyer, M. Garcia Larrode, J. Yang, A. Ng'Oma, , G.J Rijckenberg, H.P.A. Boom. "Advanced modulation techniques for polymer optical fiber transmission". proc. ECOC 07 (Th 4.1.4). (pp. 1-4). Berlin, Germany, (2007).
4. M. Haupt, C. Reinboth and U. H. P. Fischer. "Realization of an Economical Polymer Optical Fiber Demultiplexer", Photonics and Microsystems, 2006 International Students and Young Scientists Workshop, Wroclaw, (2006).
5. H.P.A.v.d.Boom, W. Li, G.D. Khoe. "CWDM Technology for Polymer Optical Fiber Networks". Proceedings Symposium IEEE/LEOS Benelux Chapter, Delft, The Netherlnads, (2000).
6. M. Vieira, A. Fantoni, M. Fernandes, P. Louro, G. Lavareda and C.N. Carvalho, Thin Solid Films, 515, Issue 19, 7566-7570, (2007).
7 M. Vieira, M. Fernandes, J. Martins, P. Louro, R. Schwarz, and M. Schubert, IEEE Sensor Journal, 1, 158-167, (2001).
8. E. S. Yang, "Microlectronic devices", Chap. 5, Department of Electrical Engineering, Colombia University, Mc Graw-Hill, Inc. (1988).

Mater. Res. Soc. Symp. Proc. Vol. 1076 © 2008 Materials Research Society

Tailoring Quantum Dot Saturable Absorber Mirrors for Ultra-Short Pulse Generation

Matthew Lumb[1], Edmund Clarke[1], Dominic Farrell[2], Michael Damzen[2], and Ray Murray[1]

[1]EXSS Physics, Imperial College London, The Blackett Laboratory, Imperial College, Prince Consort Road, London, SW7 2BZ, United Kingdom

[2]Photonics, Imperial College London, The Blackett Laboratory, Imperial College, Prince Consort Road, London, SW7 2BZ, United Kingdom

ABSTRACT

We have designed and grown a series of quantum dot semiconductor saturable absorber mirrors (QD-SESAMs) for a range of operating wavelengths, incorporating innovative design and processing features to optimise the device performance. Using a range of reflectivity studies, ellipsometric measurements and both time-integrated and time-resolved spectroscopic studies, we have conducted detailed investigations of device performance. Extensive modelling work of dielectric multilayers has been undertaken which supports our experimental findings and allows us to understand and design novel structures in order to improve and tailor device characteristics, including dielectric capping and non-normal incidence. We demonstrate samples designed for operation with the higher excited-states of the QDs which produced a self-starting train of mode-locked pulses with a temporal duration of 200 ps at a repetition rate of 78 MHz in a Nd:YVO$_4$ solid-state laser. We also present SESAMs incorporating electronically coupled QD bilayers, allowing long wavelength operation.

INTRODUCTION

Semiconductor saturable-absorber mirrors (SESAMs) are well suited as passive mode-locking elements in lasers due to their ultrafast carrier dynamics, high spectral tunability and low toxicity compared with dye-based absorbers. A SESAM consists of a multilayer mirror with an absorbing layer in the cavity region between the distributed Bragg reflector (DBR) and the surface. The entire structure can be monolithically grown by molecular beam epitaxy (MBE) and the absorber regions are usually positioned at the anti-nodes of the field distribution at the design wavelength of the sample for maximum effectiveness. The absorber is usually a quantum well (QW) whose absorption band edge can be tuned to the design wavelength of the SESAM.

Recently, quantum dots (QDs) have been used as the absorber region in SESAM structures at a range of wavelengths[1,2,3,4]. QDs have numerous advantages over QWs for use in SESAMs operating in the near-IR. They have ultra-fast absorption recovery times[5] and large inhomogeneous line-widths relative to QWs. This makes them ideal candidates for laser systems where large absorption bandwidths are required. The emission wavelength of the ground-state (GS) quantum dot transition using single QD layers capped with GaAs ranges from <1μm to around 1300nm. We have also extended the emission wavelengths to >1500nm using QD bilayers[6]. Therefore QD-SESAMs have the potential to operate with a very wide range of gain-media out to the telecommunications wavelengths.

There are several key parameters which govern a SESAM's performance and which must be suited to a particular laser system in order to work effectively. The modulation depth of the SESAM must produce sufficient non-linearity without introducing losses that are too large to sustain laser operation. Also, the saturation intensity, I_{sat}, must be well suited to the intensity in the laser cavity to optimise the non-linearity and avoid Q-switching instabilities[7]. The absorption recovery time should also be well matched to the laser system. To achieve ultra-short pulses, fast absorption-recovery times are desirable but this increases the saturation intensity[8]. Additional considerations include reducing non-saturable losses, such as, scattering from interfaces and crystal defects, to a minimum and also controlling the group delay dispersion (GDD) of the device to keep the overall cavity dispersion to a minimum. Kopf *et al.*[9] have demonstrated how SESAM design can be used to control saturation intensity and dispersion.

EXPERIMENT

All the samples were grown using solid source MBE. The dots were characterised using normal incidence and edge emitting photoluminescence (PL) measurements at low and high intensity. Low intensity PL was obtained from samples with excitation from a 5 mW HeNe laser. High intensity PL measurements were made using a 500 mW Ti:Sapphire laser operating at 790 nm. The PL signal was dispersed using a monochromator and detected by a Ge photodiode using standard lock-in techniques. Time-resolved photoluminescence measurements (TRPL) were also made on some of the samples using 2.3 ps pulses from a Ti:Sapphire laser operating at 790 nm. PL decays were obtained with a time-correlated single-photon counting system using a Hamamatsu micro-channel plate with a temporal resolution of ≈ 30 ps. The luminescence lifetimes are obtained using commercial fitting software. On samples designed for operation at 1064 nm, saturable absorption measurements were made by measuring the incident and reflected pulse fluence from an Yb-doped fibre laser, using a knife edge as an attenuator. The sample was then incorporated into a Nd:YVO$_4$ laser cavity in bounce geometry. The setup of the laser cavity is described in detail in previous work[10]. The normal-incidence reflectivity measurements were obtained using a tungsten light source and microscope, and detected with an optical spectrum analyser (OSA). A high quality Au mirror was used to calibrate the reflectivity. Layer thicknesses in the structures were determined using ellipsometric techniques and fitting to the data using commercial software. The multilayer structures were modelled using a standard transfer matrix technique first formulated by Abeles[11].

DISCUSSION

Device Design, Dielectric Capping and Non-normal Incidence

The device should have a high reflectivity across the full operating bandwidth of the device and a suitable enhancement factor, ξ. This is a measure of the field intensity at the absorber region relative to that outside the sample, and is important for determining the saturation intensity. It is also advantageous to keep group-delay dispersion (GDD) low across the stop-band region to allow the production of ultra-short pulses and reduce the necessity for dispersion compensation techniques. The index profile and the variation in the electric field intensity across the structure are shown in figure 1 for a low-finesse (LF), resonant SESAM consisting of a 25 period DBR and GaAs cavity. The dot layers are positioned at the anti-nodes of the field in the cavity region which are spaced at $\lambda_o/2$ intervals, where λ_o is the design wavelength (1290 nm in this case). In the analysis, the presence of the dot layers is neglected under the assumptions made by Spuhler[12].

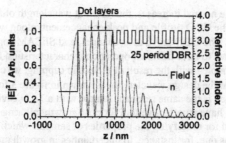

Figure 1. The refractive index and electric field structure for the resonant, low-finesse QD-SESAM designed for operation at 1290nm.

Post-growth deposition of optimal thicknesses of dielectric layers such as SiO_2 can reduce the effects of dispersion whilst retaining a large value of ξ. Deposition of a SiO_2 layer less than quarter-wave layer thickness suppresses the variation in GDD across the stop-band region. For a cap thickness of $\lambda_0/4$, the resonance condition of the sample changes from resonant to anti-resonant. Deposition of the cap also changes the effective length of the cavity and therefore the position of the cavity resonance, shifting the GDD profile. This means that the value of GDD at λ_0 can be tailored simply by deposition of a dielectric cap, as shown in figure 2. The enhancement factor is also reduced when a quarter-wave cap is deposited, but due to the index contrast of the GaAs cavity and the SiO_2 layer, the enhancement factor is not attenuated to the level of an anti-resonantly designed sample, also shown in figure 2. The resonant SESAM modelled here has a cavity thickness of $7\lambda_0/4$ and the anti-resonant SESAM has a $6\lambda_0/4$ cavity. Note that deposition of the cap has negligible effect on the reflectivity. However, capping the sample to optimise one of these parameters unavoidably impacts on the other. In order to have more control over these properties simultaneously we now turn our attention to SESAMs at non-normal incidence which impacts on the reflectivity, GDD and enhancement factor for the structure.

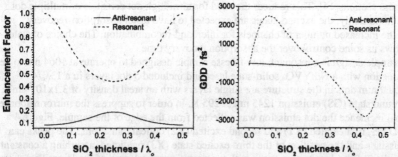

Figure 2. The enhancement and GDD at λ_0 for a range of SiO_2 cap thicknesses up to $0.5\lambda_0$ for a resonant LF-SESAM with a $7\lambda_0/4$ cavity and an anti-resonant LF-SESAM with a $6\lambda_0/4$ cavity.

As the angle to the normal increases, the resonant wavelength of the structure blue-shifts. This results in a shift of the GDD profile and peak enhancement, along with the stop-band region of the reflectivity profile. The GDD and ξ at λ_o changes, and SESAMs operating at non-normal incidence have been shown to have different operational characteristics[13]. If the non-normal incidence behaviour is coupled with the result of dielectric capping, we can produce a map of the variation of ξ and GDD at the design wavelength, as shown in figure 3 for the case of p-polarised light. The sample modelled is the same resonant sample with a $7\lambda_o/4$ cavity as shown in figure 2. From such maps we now have an effective tool for post-growth tailoring of SESAM properties, which is especially useful for modifying the properties of samples which differ markedly from their design specifications due, for instance, to uncertainties in growth rates.

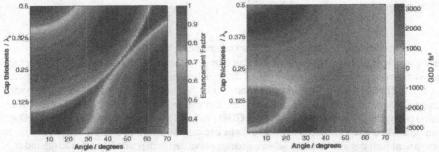

Figure 3. Maps of enhancement factor and GDD at λ_o for the resonant LF-SESAM at a range of cap thicknesses and angles.

Excited State QD-SESAMs

The presence of a number of confined states in QDs results in several inhomogeneously broadened optical transitions and the choice of dot state resonant with λ_o offers some potential advantages for the SESAM properties. Excited states have higher degeneracy and larger absorption cross sections[14]. This reduces the need for ultra-high dot densities or multiple dot layers. More importantly, the excited-states are expected to have faster absorption-recovery times due to the increased number of channels for interband-recombination. The choice of state therefore allows us some control over the absorption recovery time.

The sample design is a resonant, low-finesse sample designed to operate at 1064 nm to be used in conjunction with a Nd:YVO$_4$ solid-state laser and included 5 dot layers in a $13\lambda_o/4$ cavity. The quantum dots in the structure are single layers with an areal density of 3.1×10^{10} cm^{-2}, with peak ground state (GS) emission 1245 nm at 295 K. In order to suppress the mirror response seen at normal incidence the dot emission was collected from the edge of the sample. Figure 4 shows the QD GS, first excited (X1) and second excited (X2) optical transitions. The peaks can be fit by Gaussians and the position of the third excited state (X3) predicted, assuming a constant energy separation[14] is close to resonance with the operating wavelength of the laser. The modulation depth and saturation fluence, measured from saturable absorption measurements, were found to be 0.92% and 97.7±3.3μJ/cm^2 respectively. When placed in the laser cavity, the sample produced a self-starting train of mode-locked pulses with a temporal duration of 200 ps at a repetition rate of 78 MHz (see inset of figure 4). To our knowledge this is the first demonstration of CW mode-locking from a QD-SESAM operating in an excited state.

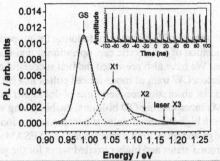

Figure 4. The edge-emitted PL from the 1064 nm QD-SESAM under high power excitation. **Inset.** Self-starting, CW mode-locked pulses using the SESAM in a Nd:YVO₄ laser.

Work is currently underway to investigate SESAMs for long wavelength operation. Two test SESAMs for 1290 nm have been grown to a design identical to that shown in figure 1. Sample A is designed to have the GS transition resonant with λ_o whilst sample B has the X1 state resonant. The dot layers in sample A were single-layer dots whereas the dot layers for sample B are bi-layers[6]. Figure 5a compares the PL obtained from dot samples grown to the same recipe, demonstrating the resonance of the GS and X1 states of the dots with the design wavelength of 1290 nm, through precise control of the MBE growth conditions. The inset shows that the lifetime of sample B (516±4 ps) is significantly shorter than that of sample A (689±4 ps) at the target wavelength, consistent with the notion of a faster absorption recovery time for the X1 state compared with the GS. Figure 5b shows that the normal incidence reflectivity of sample A is in good agreement with the calculated reflectivity with very little non-saturable loss. The layer thicknesses are calibrated using ellipsometric measurements. This preliminary work shows that use of QD bi-layers enables QD-SESAM operation to be extended out to telecommunications wavelengths and furthermore enables long wavelength operation in dot excited states, which makes QD-SESAMs very promising solution for generation of ultra-short pulses at telcommunications wavelengths.

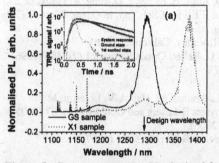 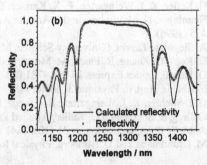

Figure 5a. Room temperature PL obtained from samples A and B. **Inset:** room temperature TRPL data for the two samples showing the faster lifetime of the X1 state. **Figure 5b.** Measured normal incidence reflectivity and the calculated reflectivity and GDD profiles for sample A.

CONCLUSIONS

We have discussed a range of techniques for tailoring the properties of QD-SESAMs, including post growth deposition of a dielectric cap, operation at non-normal incidence and use of dot higher excited states. We have also presented the first successful operation of an excited-state QD-SESAM to produce a CW train of mode-locked pulses in a Nd:YVO$_4$ laser, exploiting the higher degeneracy and fast absorption recovery time of dot higher excited states. We have also demonstrated SESAMs incorporating QD bi-layers, enabling long wavelength operation and operation of QD-SESAMs in higher excited states at telecommunication wavelengths. Work is currently underway to design and grow highly specialized QD-SESAMs for use in a wide variety of laser systems in both ground states and higher excited states for the generation of ultra-short pulses.

ACKNOWLEDGMENTS

The authors would like to thank Andrei Rulkov and Burly Cumberland for their useful help and advice. This work was supported by EPSRC, UK.

REFERENCES

1. E. U. Rafailov, S. J. White, A. A. Lagatsky, A. Miller, W. Sibbett, D. A. Livshits, A. E. Zhukov, and V. M. Ustinov, Photonics Technology Letters, IEEE **16** (11), 2439 (2004).
2. K. W. Su, H. C. Lai, A. Li, Y. F. Chen, and K. E. Huang, Optics Letters **30** (12), 1482 (2005).
3. A. A. Lagatsky, F. M. Bain, C. T. A. Brown, W. Sibbett, D. A. Livshits, G. Erbert, and E. U. Rafailov, Applied Physics Letters **91** (23) (2007).
4. C. Scurtescu, Applied physics. B, Lasers and Optics **87** (4), 671 (2007).
5. P. Borri, S. Schneider, W. Langbein, U. Woggon, A. E. Zhukov, V. M. Ustinov, N. N. Ledentsov, Z. I. Alferov, D. Ouyang, and D. Bimberg, Applied Physics Letters **79** (16), 2633 (2001).
6. E. C. Le Ru, P. Howe, T. S. Jones, and R. Murray, Physical Review B **67** (16) (2003).
7. U. Keller, K. J. Weingarten, F. X. Kartner, D. Kopf, B. Braun, I. D. Jung, R. Fluck, C. Honninger, N. Matuschek, and J. Aus der Au, IEEE J. of Sel. Top. in Quant. Elect., **2** (3), 435 (1996).
8. A. Siegman, *Lasers*. (University Science, Mill Valley, California, 1986).
9. D. Kopf, G. Zhang, R. Fluck, M. Moser, and U. Keller, Optics Letters **21** (7), 486 (1996).
10. D. Farrell, Optics Express **15** (8), 4781 (2007).
11. F. Abeles, Ann. d. Physique **3**, 504 (1948).
12. G. J. Spuhler, K. J. Weingarten, R. Grange, L. Krainer, M. Haiml, V. Liverini, M. Golling, S. Schon, and U. Keller, Applied Physics B-Lasers and Optics **81** (1), 27 (2005).
13. A. McWilliam, Optics letters **31** (10), 1444 (2006).
14. M. Grundmann and D. Bimberg, Physical Review B **55** (15), 9740 (1997).

AUTHOR INDEX

Abedin, M. Nurul, 147
Adany, Peter, 57
Albert, Jacques, 65
Allen, Chris, 57
Allison, Steve, 181
Ambacher, Oliver, 155
Amzajerdian, Farzin, 3, 47, 147
Ando, Toshiyuki, 35
Asaka, Kimio, 35

Bai, Yanbo, 81
Bandara, Sumith, 133
Banerjee, Koushik, 127
Barata, Manuel, 187
Barlow, Fred, 119
Barnes, Bruce, 47
Blazejewski, Edward, 133
Büchner, Hans-Joachim, 155

Clarke, Edmund, 193

Damzen, Michael, 193

Elshabini, Aicha, 119

Fabre, Frédéric, 9
Fallahi, Mahmoud, 91
Fan, Li, 91
Fantoni, Alessandro, 187
Farrell, Dominic, 193
Feeler, Ryan, 119
Fernandes, Miguel, 187

Garnweitner, Georg, 175
Ghosh, Siddhartha, 127
Grein, Christoph, 127
Gunapala, Sarath, 133

Hader, Joerg, 91, 97
Hansel, Rachael, 181
Hauguth-Frank, Sindy, 155
Hessenius, Chris, 91
Hill, Cory, 133
Hirano, Yoshihito, 35
Hui, Rongqing, 57

Inokuchi, Hamaki, 35

Jäger, Gerd, 155
Junghans, Jeremy, 119

Kameyama, Shumpei, 35
Kemner, Greg, 119
Keo, Sam, 133
Koch, Stephan W., 91, 97
Krishna, Sanjay, 127
Kups, Thomas, 155

Lebedev, Vadim, 155
Li, Hongbo, 91
Li, Li, 65
Lian, Cheng-Wei, 165
Lin, Gong-Ru, 165
Liu, Jian, 75
Liu, John, 133
Lockard, George, 47
Louro, Paula, 187
Lumb, Matthew, 193

Mallick, Shubhrangshu, 127
Moloney, Jerome, 91, 97
Morancais, Didier, 9
Mumolo, Jason, 133
Murray, James, 91
Murray, Ray, 193

Novo-Gradac, Anne-Marie D., 27

Petway, Larry, 47
Peyghambarian, Nasser, 65
Pierrottet, Diego F., 47
Plis, Elena, 127

Razeghi, Manijeh, 81
Refaat, Tamer F., 147
Rodriguez, Jean Baptiste, 127
Romanus, Henry, 155
Rubio, Manuel, 47

Schober, Andreas, 155
Schulzgen, Axel, 65

Shaw, George B., 27
Slivken, Steven, 81
Stephens, Edward, 119
Stolz, Wolfgang, 91
Sulima, Oleg V., 147
Suzuki, Shigeru, 65

Tanaka, Hisamichi, 35
Temyanko, Valery L., 65
Ting, David, 133
Tonisch, Katja, 155
Toulemont, Yves, 9
Tsedev, Ninjbadgar, 175

Vaseashta, A., 21
Vieira, Manuel Augusto, 187
Vieira, Manuela, 187

Walker, Greg, 181
Wood, Jared, 119

Yu, Anthony W., 27

Zhang, Wei, 81
Zhu, Xiushan, 65

SUBJECT INDEX

absorption, 193

Be, 27

ceramic, 119
composite, 175
crystalline, 181

devices, 127, 133, 147
dopant, 181

electron-phonon interactions, 21
electronic material, 147
epitaxy, 97

fiber, 35, 57, 65

government policy and funding, 3

holography, 175

infrared (IR) spectroscopy, 133

laser, 3, 9, 27, 35, 47, 57, 65, 75,
      81, 91, 119, 165, 193
luminescence, 97

microelectronics, 187
microstructure, 65

nanostructure, 21, 165, 175

optical, 9, 27, 57, 75, 81
   properties, 75, 91, 97, 193
optoelectronic, 9, 47, 127, 155, 187

photoconductivity, 133, 187

radiation effects, 3

sensor, 21, 35, 47, 181
Si, 165
structural, 155

III-V, 81, 91, 119, 127, 147, 155

Printed in the United States
By Bookmasters